Crystals and Crystal Growing

"I would remind you then that when we want to find out anything that we do not know, there are two ways of proceeding. We may either ask somebody else who does know, or read what the most learned men have written about it, which is a very good plan if anybody happens to be able to answer our question; or else we may adopt the other plan, and by arranging an experiment, try for ourselves."

C. V. Boys, *Soap Bubbles and the Forces Which Mould Them* (A Science Study Series book)

Crystals and Crystal Growing

Alan Holden and Phylis Morrison

Introduction by Philip Morrison

The MIT Press
Cambridge, Massachusetts
London, England

First MIT Press edition, 1982

New material copyright © 1982 by The Massachusetts Institute of Technology

Copyright © 1960 by Alan Holden and Phylis Morrison

This book was printed and bound in the United States of America.

Library of Congress Cataloging in Publication Data

Holden, Alan.
    Crystals and crystal growing.

    Includes bibliographical references and index.
    1. Crystals.   2. Crystals—Growth.
I. Morrison, Phylis, 1927-   II. Title.
QD921.H58   1982   548   81-23639
ISBN 0-262-58050-0 (pbk.)   AACR2

INTRODUCTION
TO THE MIT PRESS EDITION

Once upon a time. . . . It seems natural to use the phrase for the almost-mythical past to evoke the time when this book was first brought out, in 1960. It is one of a long series of books then "published as part of a fresh approach to the teaching and study of physics." A group of enthusiastic people did the job. They were university physicists and high school teachers, journalists and film producers, apparatus designers and other specialists, even a few high school students, organized as the Physical Science Study Committee, who first assembled at MIT in 1956. Their task was the construction of a new and many-sided path to a physics course for high school students, with a new textbook, a lab guide with especially designed apparatus to make it live, an extensive series of films, a source book for teachers and plenty of opportunities for teachers to learn, to try out, and to evaluate the new materials. Part of the task included the stimulation of a large choice of volumes to be used as supplements to the work in class, through which authors who knew something of value and wanted to express it were given a chance to speak to interested students and teachers.

Those books covered a very wide range of topics, just as physics does. The MIT Press has already brought out (in 1979) one of the most delightful of all the books of the series, the fine *Knowledge and Wonder*, by MIT theoretical physicist Victor Weisskopf. That book takes an insightful and wide-ranging, almost philosophical look at the full spread of natural phenomena and what understanding them means to human beings. It begins with galaxies and ends with life and evolution, taking a climb along the whole quantum ladder. The book at hand lies at almost the other end of the spectrum of exposition in physics. It takes careful, narrow focus on growing your own marvelous big gem-like crystals in a couple of Mason jars standing in some undisturbed corner of your home. That task will hardly succeed by slavish following of some mysterious recipe. But by learning to apply a simple but ingeniously reasoned strategy, molecules can be coaxed into attaining all the order their nature allows. Along the way the book necessarily deals with the qualities of order and disorder in nature, in particular in the crystalline state, whose order is so hidden in daily life, yet so plain in the form and elegance of a glistening crystal big enough to hold in the fingers.

It was a hot summer morning when the coauthors brought me at MIT a cigar box full of such crystals, grown in and around the classroom. Their beauty and meaning were a revelation and a surprise. For just then I was writing the chapter of the physics text that dealt with atoms and molecules. It was a sore temptation to talk about growth of crystals, for their visible order is an old path to the understanding of the atomic substructure of the material world. But all the experienced hands I had asked were skeptical; you can't grow good crystals without specialists and fancy labs, they thought. How wrong they were is proved by this book, which goes on to discuss not only crystal-bearing solutions and how they work, but also the engaging design within crystal symmetry, and the revelation of that deep-lying symmetry by eye and knife blade, Polaroid sheet, hammer and neon lamp, and a piece of plastic diffraction grating.

It is easier to read and argue out these paper pages than to grow the real thing. For it takes care, patience, and understanding to make the real world follow any path made clear in words. It is safer, too, for experience has shown that those who succeed in growing their own crystals lie in real danger of spending the rest of a lifetime in some career pursuing the kind of order and disorder that lie in science and technology! With that fair warning, I endorse this book, in admiration and pleasure and a little honest envy.

Twenty years have changed the world a lot; crystals are everywhere concealed, big, perfect ones of silicon, sawed thin and made into little chips within just about everything electronic. The world of order and disorder they come from remains what it was, as the authors describe it, and the road they mapped into it is still the most inviting route I know. That energetic and insightful MIT physicist, Jerrold Zacharias, whose leadership brought us all to work together in those years, will happily approve of this new MIT Press edition; and all of us would like once again to evoke the memory of Francis Friedman, tastemaker of the PSSC, whose sure touch warranted the quality of all its products.

Philip Morrison
Cambridge, Massachusetts

CONTENTS

PREFACE

The beauty of large single crystals is arresting. The flatness of their faces, the sharpness of their angles, the purity of their colors will give you deep satisfaction.

But along with the sense of delight, you will surely have a sense of wonder. In this book, we ask you to indulge that sense also, and we suggest how you might go about it. You will find that it takes patience, care, thoughtfulness, and some feeling for the route you are traveling, together with some open-mindedness about where and when you will arrive. Adding these things successfully to your curiosity, you become a scientist.

It is a pity that most people think a scientist is a specialized person in a special situation, like a lawyer or a diplomat. To practice law, you must be admitted to the bar. To practice diplomacy, you must be admitted to the Department of State. To practice science, you need only curiosity, patience, thoughtfulness, and time.

The scientist's effort to understand the workings of the world has two sides. On the one hand, he performs experiments on bits of the world, to find out how those bits behave. He makes the assumption that another bit of the world, similar to the one he examines, ought to behave in the same way.

For this reason he demands that a valid experiment shall be "repeatable": anyone else, told how to perform the same experiment, must be able to repeat it in any other part of the world and get the same result. This demand distinguishes his experiments from those of people who claim to tip tables without touching them, to see the future before it occurs, or to transfer thoughts into other persons' minds without using the usual means of communication. So far, those people have not described their experiments in such a way that others can repeat them.

On the other hand, the scientist's activities also include a great deal of thinking, and of visualizing things in his mind's eye. While he is engaging in this side of his work, he is not looking at the real world as directly as he does in his experiments. He is trying to bring the experiments, done either by himself or by others, into some kind of orderly relationship—to derive from them a way of thinking that includes them all.

Out of these efforts come "pictures of the facts" and "laws of nature": ways of organizing physical experiences in the mind so that they can all be thought about together, as if they were related experiences. It is only when he has succeeded in finding such organizing principles that the scientist feels satisfied, happy, and ready to move on to another problem.

Here again the scientist makes his demand that an experiment be repeatable. He requires that his visualizations and thoughts should satisfy others as well as himself. Of a visualization, others must say, "Yes, that seems reasonable." Of a thought, they must say, "Yes, that is logical." This requirement distinguishes his thinking from that of "crackpots" who, as Albert Einstein once wrote, "place the existing science in denial."

Indeed the inclusive visualization, the reasonable simplification, the logical deduction, are the ingredients of both truth and beauty in the scientist's inner world. You may even hear him say of some physical theory, "The theory is so beautiful that it must be true."

For both his experimenting and his thinking, the

scientist's ways of training himself are very similar. In order to acquire a clean, quick, reliable laboratory technique, he practices performing experiments which others have done before him. When he can reproduce the results obtained by more experienced workers, he is ready to perform experiments not previously attempted. Similarly he practices thinking: In order to develop sharp, dependable intellectual tools, he works problems whose answers have been obtained by others. When he can duplicate some of the thinking done by his predecessors, he is ready to think for himself.

This book suggests many experiments which will help you to develop laboratory skill, if you want to. It also describes some of the ways you can visualize the results of these experiments, and place them in a satisfactory framework of understanding. To help you to sharpen your intellectual tools, it offers problems, in visualization and thought, which you can answer with the material presented. We urge you to grow the crystals and build the simple models and apparatus you will read about; you will find them of great help in your understanding of the subject matter.

PREFACE
TO THE MIT PRESS EDITION

It is gratifying to find that after twenty years this little book still serves a useful educational purpose. Now as then, it sets for itself three major goals: 1. Describing the atomistic character of crystallinity; 2. Describing some techniques for preparing large single crystals; and 3. Describing some experiments that display the unexpected properties which flow from crystallinity.

We have chosen not to "modernize" the text, feeling that the text of any book mirrors something of interest about the time when the words are set down. Our new readers may notice differences from the world of 1960; changes in familiar materials, in technology, even changes in the way things are described. Except for the TV tube, vacuum tubes have disappeared, to be replaced by the now familiar solid state technology of the transistor and the chip—all crystalline, with controlled defects. Nor do we imagine that *glass* would be described as we did by someone writing today. However, given the chance, atoms will still line up in parade-ground arrays, and those arrays will still have the properties endowed by that geometric order. We hope that readers of this book will explore for themselves in that surprising and beautiful world.

New Vernon, New Jersey
Cambridge, Massachusetts
November 1981

CHAPTER I

Solids and Crystals

Are glaciers made of quartz? Several centuries ago people thought they were. But today you know better. Glaciers are made of ice.

Here is an interesting instance of what a word can do to confuse matters. The word "crystal" comes from Greek roots meaning "clear ice." Who now would deny that quartz is "crystal" and glaciers are "clear ice"?

Even today the word "crystal" is still a confusing one; an unabridged dictionary will show you many different uses of it. Is a crystal a clear ball in which to gaze at the future? A cut-glass punch bowl? A gem set in a ring? Has it plane faces and sharp angles? Can you see through it? In one definition or another the dictionary will probably say yes.

But to these questions a physicist will answer, "Maybe; it all depends." He has a definite use for the word "crystal"—one of many uses given in the dictionary—which he shares with the chemist, the metallurgist, and the crystallographer. The shortest definition he could give you would be, "Crystals are solid, and solids are crystals," not a very helpful definition until he expands it. And "solid"? He has a definite use for that word too.

To your question about the punch bowl, he may give
you the answer, "No, that isn't crystal, because it isn't
solid." It is a surprising answer, perhaps even an an-
noying one to anybody who has broken a punch bowl
and found that the fragments of glass are "solid" enough
to cut him. But be patient with the physicist and hear
his reasons. You will get new insights into the world
of matter; you will learn to see his crystals in the most
unexpected places. And in later chapters, you will learn
to make some fascinating crystals at home yourself, and
to perform revealing experiments with them.

The Three States of Matter

Since the physicist says, "Crystals are solid," look first
at the familiar word "solid." One physicist has written
that solids are those parts of the material world which
support when sat on, which hurt when kicked, which
kill when shot. Roughly, but graphically, he has distinguished solids from fluids. Fluids include both liquids
and gases, and indeed another physicist has said, "We
distinguish three *states of matter:* solid, liquid, and gas."
Following this clue, think critically for a moment about
those three familiar states.

Some materials can take any of the three forms, with
no change in their chemical composition. Steam, water,
and ice are common names for the three forms taken by
a single material. Another familiar liquid, one of the
most convenient to use in making thermometers, is the
metal mercury. It is a "liquid metal." But it will freeze
to a solid metal at a low enough temperature and then,
as long as you keep it cold enough, it will behave much
like the more familiar solid metals. At a high enough
temperature, mercury will vaporize to a gaseous metal;
it forms the gas in mercury-vapor lamps, for example.

Bubbles of carbon dioxide fizz out of beer and soft

drinks, showing that carbon dioxide is a gas under ordinary conditions. At a low enough temperature the same material forms the solid called "dry ice." Unlike ordinary ice, that "ice" is "dry" because the liquid form of carbon dioxide never appears at ordinary pressure. The material disappears into thin air when you put it in a warm place because it turns directly from a solid into a gas. But you can make liquid carbon dioxide if you seal the stuff up and put it under a higher pressure.

Even common salt or a quartz rock will melt into a liquid if you get it hot enough. In fact, if you get these things even hotter than that, you can make gases out of them.

But you cannot make liquids and gases of all solids —not even out of most of them. If you heat gunpowder, it goes off with a bang, and you might say you have turned it into gas. But the change of solid gunpowder into "gunpowder gas" is not the same kind of change as that of ice into steam: you cannot get the gunpowder back by cooling the gas, as you can get ice by cooling steam. The change in gunpowder is sometimes called a "chemical" change, and the change in ice a "physical" change.

If you heat sugar very carefully, it melts into a clear liquid. But if you heat it too much, it turns brown, and you do not get all the sugar back when you cool it. The sugar decomposes chemically into new materials, which are sometimes collectively called "caramel." If you try to boil the sugar to form a gas, it will decompose completely, giving off steam and leaving a crust of charcoal.

In their ability to stand heat, most solids are more nearly like sugar than salt: they decompose chemically at temperatures below their melting points. This is the reason why vastly more materials are known in the form of solids than in the form of liquids or gases.

The best way to picture the difference in the three

states of matter is to think about a material that can appear in any one of them—mercury, for example. Picture what happens when you cool it from the form of a gas, first, into a liquid and, finally, into a solid.

In the gas the mercury atoms seem to be taking up quite a lot of space. An amount of mercury gas of the same volume as a drop of mercury, does not weigh nearly as much as the drop because there is very much less matter in it. Another way of looking at the difference is to notice that if you turned a drop of liquid mercury into mercury gas at the same pressure and temperature, the gas would occupy about a thousand times as much space as the drop did. Figure 1 contrasts the volumes of liquid and gas.

You might think that each atom of the mercury in the gas had swelled up like a balloon and was a thousand times bigger in the gas than it was in the liquid. But that is not so; the atoms stay nearly the same size. They use up more room by dashing about. As they travel, they bounce off each other and off the walls of their container. Each bounce off the wall gives the wall a little kick, and the combination of all these little kicks is the pressure of the gas on the wall.

But the important thing to notice is that the gas is a completely disorderly collection of mercury atoms. Indeed, it is disorderly in two distinguishable ways. Except during a collision, no atom pays any attention to what the speed and direction of any other atom may happen to be. The atoms are moving about quite independently, except during the instants when two collide, and thus their *motions* are entirely disorderly. And if you could take a quick picture of the collection at any instant, you would see no pattern in the picture: in other words, the *positions* of the atoms are entirely disorderly too.

To keep the gas together in one place, you must confine it, so that the atoms do not speed off into space in all directions. If you cool the confined gas, the atoms move less rapidly. And finally, at some lower temperature, the activity of the atoms slacks off enough to give

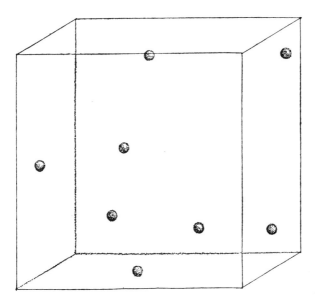

Fig. 1. A COMPARISON OF VOLUMES *shows that the atoms in a gas are far from one another and hence can move independently; the atoms in a liquid or a solid are closely packed and restrict one another's motions. At the normal temperature and pressure of the atmosphere a given number of atoms in a gas occupies about a thousand times the volume—ten times the linear dimension—of the same number of atoms in liquid or solid form.*

the *attractions* of the atoms for one another a chance to have a big effect. Then the attractions bring most of the atoms together into a liquid, and drops of mercury form in the container.

Not all the mercury gas collapses in this way; there is still a gas of mercury atoms in the container. If you remove the confining walls, that gas will escape, and mercury atoms will leave the surfaces of the drops in an effort to regenerate the lost gas. In other words, the mercury will slowly evaporate.

The evaporation shows that the mercury atoms are still moving about in the liquid. Even when the confining walls are there, mercury atoms are leaving the liquid to join the gas, but just as many atoms are leaving the gas to join the liquid. Such a balance of escape and capture is often called "dynamical equilibrium." In short, both the mercury gas and the mercury drop, apparently stationary and unchanging after equilibrium has been established in the container, are really the seats of a great deal of activity.

Look for a moment at the pool of liquid mercury which forms as the drops gather together. Since the liquid flows freely, the atoms must be able to move past one another quite easily. But it holds together in drops if you spill it, showing that there are attractive forces between the atoms—the same forces which brought them together to form the liquid from the gas, and which now keep them associated in densely packed liquid communities.

You can get a good idea of how close this association must be if you contrast the effects of pressure on the gas and on the liquid. If you apply pressure to the gas, you can easily squeeze it into a smaller volume. But even quite a high pressure does not reduce the volume of the liquid much. From this you can guess that the atoms in the liquid are packed together almost as tightly as they can be.

Of course, there is still a little free space in the liquid —space enough for the atoms to move about irregularly. And there is a constant competition between the attractive forces which hold the atoms together and the irregular motions which tend to knock them apart. The atoms evaporating from the surface, for example, are those which get a powerful enough kick, from the atoms beneath the surface, to knock them out of the liquid despite the attractive forces tending to keep them there. When you cool the liquid, the motions are less vigorous, the evaporation is less rapid, and the attractive forces can pull the atoms even closer together so that the liquid contracts slightly.

As you continue to cool the mercury, it finally reaches a temperature at which it solidifies. When that happens, there is usually very little change of volume, because the atoms are already packed together so closely in the liquid that solidification cannot bring them much closer. But the solid does not flow freely, as the liquid did, and you can guess that the atoms can no longer move past one another easily. In other words, liquid and gas are alike in their ability to flow but unlike in the volumes they occupy, while liquid and solid are alike in the volumes they occupy but unlike in their ability to flow. Figure 2 may help you to picture the differences in the structures of gas, liquid, and solid.

From the rigidity of solids, which suggests that the atoms cannot move past one another, you might jump to the conclusion that the atoms have stopped moving altogether. But that would be going too far. Remember that dry ice evaporates: the molecules of carbon dioxide escape from the solid to form carbon dioxide gas. Hung wet on the clothesline in winter, the week's wash may freeze stiff, yet it will dry out. Just as the mercury atoms leave a drop of liquid mercury when it evaporates, the molecules escape from ice because they are moving all the time, and occasionally one is kicked by its neighbors hard enough to go flying off. But the departure of mole-

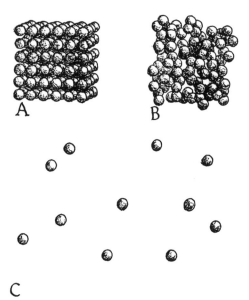

Fig. 2. THE THREE STATES OF MATTER. *A—In the solid, atoms are close together. They vibrate but cannot move past one another. B—In the liquid the atoms are almost as closely packed as in the solid, but they can move past one another. C—In the gas the atoms are widely separated, and can move almost independently.*

cules from most solids at ordinary temperature is so infrequent that the evaporation is imperceptible.

Since the atoms in a solid are moving but cannot pass one another, each atom stays in one place *on the average,* and its motion is a vibration about that place. It has a fixed average position from which it keeps making little excursions. The atoms box one another in, so to speak, and the average position of each atom is in the middle of its own private box.

Order and Disorder

The atoms in a solid exhibit another feature, a most important one. The fixed average positions, about which the atoms vibrate, are arranged in an orderly way. And the orderliness is of a particular sort: the solid consists of a pattern of atoms repeated again and again. In two dimensions you can liken the result to the repeated design of wallpaper. In three dimensions it is like a large hotel with floor upon floor of identical rooms, identically furnished. Usually any one material has only one preferred orderly arrangement, and usually different materials have different orderly arrangements.

The orderliness of solids is a rather astonishing fact of nature. Physicists have become used to the fact, and they often forget that they do not really know why atoms adopt orderly arrangements. Nevertheless, more than any other property, this orderliness distinguishes solids from liquids: the atoms are packed closely together in both, but they have a constantly shifting, disorderly arrangement in a liquid and an orderly arrangement about which they vibrate in a solid. Orderliness of this regularly repeated sort is called "crystallinity"; anything having crystallinity is a "crystal" or a collection of crystals. That includes almost all solids and it includes very little else.

You may ask, "How far must order extend to make a material solid?" Of course, an atomic hotel can be pretty large from an atom's point of view, and at the same time pretty small from a human point of view. An atom is only about one hundred millionth of an inch in diameter. If an atom were as large as a golf ball, the atoms in an inch would stretch from New York to San Francisco. There are a hundred million times a hundred million times a hundred million atoms in a cubic inch of crystal.

A millionth of an inch is a fairly long distance from an atom's viewpoint—a hundred times its own size. Certainly you can say a substance is crystalline if it has regular arrays of atoms extending over a millionth of an inch. If the component crystals in a piece of matter are that large, each crystal contains a million atoms in orderly array, but you still cannot see it through a microscope.

Thus a crystal is not necessarily a single beautiful solid with plane faces bounding it. If you break a piece from the solid, the fragment is still a crystal, because the orderly arrangement of atoms extends a "long" distance. If you pound the fragment to dust, you have many crystals, for each of the dust particles is much larger than a millionth of an inch.

All pieces of metal are crystalline, as Plate 1 suggests: they are like the crystalline dust, but perfectly dense, with no space between the individual crystals. The crystals are jumbled together every which way, but within each of them the arrangement of atoms is orderly. Making crystals like those shown in Plate 2 will give you a vivid impression of what a metallic crystal looks like when it stands alone.

If you melt the dust or the metal, however, you destroy most of the orderly arrangement. There is some fluctuating, local, temporary order in the liquid—groups of a hundred atoms become ordered for a short time perhaps—but order is to be found only here at one instant, there at the next. And when you vaporize the liquid, even those relics of orderliness disappear.

How do we know that solids are orderly? Today we have powerful methods for confirming the idea. But long before those methods were available, the idea first arose in the minds of people who looked thoughtfully at the evidence gathered by their unaided eyes. The same evidence is accessible to you.

Consider mica. Glittering flakes of it are in the rocks all about you. The flakes easily come apart into thin sheets, and the only limit to their thinness is your skill in picking at them, by techniques such as that shown in Plate 3. It is hard to avoid the conclusion that the atoms in the mica are arranged in great sheets. Then if the attractive forces between atoms in each sheet are strong, and the forces between atoms in adjacent sheets are weak, you would expect what you actually observe.

Consider salt. If you dissolve it in water and then let the water slowly evaporate, you will get back the salt in little cubes, with flat faces. You can dry one of the little cubes on a paper handkerchief and cleave it with a razor blade along planes parallel to the cube faces, making more little cubes with flat faces out of the original. It is hard to avoid the conclusion that the atoms are all arranged in some orderly way which predisposes salt to form cubes. Salt in the salt shaker, even though it has been knocked about, usually consists of little cubes, as you could see with a low-power magnifying glass (Plate 4).

If you make a crystal of alum, by the recipe given later, you will get a solid that looks like the crystal pictured in Plate 5. During its growth it retains its shape, just getting bigger. Why does it grow into that shape, all by itself, when you impose no shape on it? It is hard to avoid the conclusion that the whole crystal is made of regularly arranged little cubic units—atomic building blocks—stacked together in the way the picture shows.

Many of the arguments and experiments in later chapters will give you other evidence of orderliness in solids. The experiments show, among other things, that many

properties of crystals are somewhat like the cleavage of
mica and salt: they depend on direction, and not on the
point from which you start. It is difficult to see how a
property of a material can depend on direction unless
the material has some underlying orderliness.

Notice also that this evidence all suggests a particular
sort of orderliness: the *repetitive* sort. The sheets of mica
are all alike, whatever part of the crystal you take them
from. And the fact that properties we discuss in a later
chapter depend on direction, but not on the starting
point in the crystal, shows that the orderly arrangement
must be repeating itself within very short distances—distances comparable to the size of an atom.

Of course, many other sorts of orderliness can be imagined, but they are not repetitive and therefore not the
sorts the atoms in a crystal take up. To most people
"orderliness" means anything done according to a clearly
established rule. But there are many kinds of rule that
do not lead to a repetition of any pattern, and hence are
not possible kinds of order for the atoms in a solid.

In Figure 3A you see an orderliness that is *not* the
kind that atoms adopt in a crystal. The atomic positions
proposed there do follow a rule, and the continued application of the rule would continue to produce an orderly arrangement of a sort for the atoms. The arrangement would even give the whole piece of matter the same
density in all its parts. But the rule is not one that repeats the same pattern from place to place. On the other
hand, Figure 3B proposes a kind of order that repeats
itself again and again as you extend it.

The experiments later in the book will remind you of
something else, something you know already but may
forget because you know it so well: the delightful *diver-
sity* of solids. There are all kinds of solids—hard and
soft, clear and colored, high melting and low melting
—many more kinds than there are kinds of gases. The

only way in which gases differ is in the kinds of molecules they are made of. In other ways all gases seem to be much alike; any gas will mix completely with any other gas, for example. Liquids show more differences, to be sure; oil floats on water without dissolving in it. But solids have endless variety.

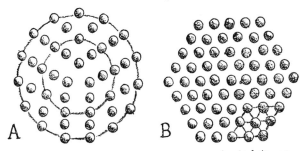

Fig. 3. CONTRASTING KINDS OF ORDER. *A—At left proposed sites for atoms are located by a rule. Successively more atoms are placed on successively larger circles. The result is an orderly placement of sites without a repeated pattern and hence not "crystalline." B— At right, the sites are placed on the intersections of a regular triangular net of lines. They form a repeated pattern; their repetitive order is "crystalline" in the sense that one can start from any corresponding point and see the same order.*

One reason for the variety of solids is the fact that more substances are chemically stable in solid form, as you noticed at the beginning of this chapter. But another reason is that solids are orderly. Solids differ not only in the kind of molecules they are made of, as gases do, but also in the kind of orderliness those molecules adopt in the solid. At first you might think that there were more kinds of disorder than kinds of order, but a second look will convince you that there is really only *one* kind of disorder for things as small and numerous as atoms.

The patterns in Figure 4 will help you to take that

second look. They are made of two motifs, strewn in disorder at A and arranged in three different orderly patterns at B, C, and D. You may say, "The diagram at A is just as orderly as the rest; it is only more complicated." But you must contemplate the problem of continuing it in a disorderly way.

If you continue it by repeating it, you construct a form of orderliness: the orderly repetition of that pattern. You

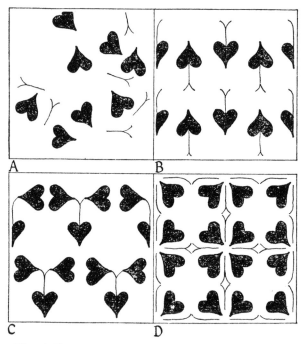

Fig. 4. FOUR DIFFERENT ARRANGEMENTS *of the same motifs. There is only one kind of disorder (A), because random repetitions of the motifs cannot be told apart. On the other hand, there are many distinguishable ways of putting the motifs in repetitive order, such as are shown at B, C, and D.*

can continue it in a disorderly way only by making each added pattern different from every other. As you do this, you use up more and more of the possible ways of strewing the motifs in disorder within each pattern. In the resulting construction all the patterns would be different and placed next to one another at random.

Hence, any such effort will look like any other such effort. You cannot distinguish two different disorderly constructions in any other way than by specifying exactly where each item is located in each construction. When the items are as small and as numerous as the atoms in a solid, you have no way to see the exact specifications of a construction, and consequently all the constructions look alike.

A crowd of people at a college football game will often remind you that disorder is a single thing, whereas order is many things. Before the game begins, the members of the crowd will be arranged in a disorderly way. Looking down from the top of the stadium, you cannot distinguish the different arrangements from one minute to the next. But later the members of the college bands will leave the crowd, and you will see them march in many distinguishable sorts of orderliness.

Orderliness comprises so many different possibilities that almost every substance has a kind of orderliness of its own. Only when two substances are very much alike will they have the same kind of orderliness. For this reason, when several substances are all dissolved in a single solution, they will usually crystallize out separately. Crystals of a substance will usually contain very little, if any, of the other substances present when the crystals formed. If you add sugar to the salt solution which you evaporate to get salt cubes, the salt cubes will contain almost no sugar.

There are some cases, however, in which mixtures of different substances will combine in an orderly way. The alloy of mercury and silver shown in Plate 2 is one. Mercury alone will form crystals in which the atoms are arranged in one kind of orderliness, and silver alone will form crystals with another kind of orderliness. Mercury and silver together form crystals with still different sorts of orderliness.

Alum, the crystal shown in Plate 5, is one of the most beautiful examples of these so-called "double compounds." Potassium sulfate and aluminum sulfate can be crystallized separately, into crystals with their own different kinds of atomic orderliness. But if you dissolve both substances together in water, the solution will deposit crystals of alum, potassium aluminum sulfate, whose orderliness differs from that of the crystals of either ingredient.

In such a crystal the molecules of the ingredients are always present in a simple numerical ratio. The ratio of potassium sulfate to aluminum sulfate is precisely one to one, in terms of numbers of molecules. Each substance has its own role to play in the construction, different from its partner's role. Both must co-operate; and since neither can replace the other, there is no way for the crystal to include more or less than a fixed ratio of its ingredients. In the case of mercury and silver, alloys with several different simple ratios are observed. But each ratio is fixed, and to each there corresponds a different alloy, with a different example of orderliness.

Crystal Growth

The orderliness of the atomic arrangement in a crystal is certainly its most important feature. But another feature, almost as important, is the fact that a crystal does not suddenly spring into being; it *grows* into being. If a solid is made of crystals—and the physicist says that most solids are made so—then you can understand solids only by understanding this aspect of crystals too. The fact that a crystal must grow may seem obvious, but once you get that fact out in the open and look at its consequences, you will quickly see that it explains many conspicuous properties of solids.

In the first place, a growing crystal clearly does not draw its nourishment from within. It has to grow from outside—from the stuff presented to its surface. The stuff must be the right stuff, able to accommodate itself to the particular kind of orderliness possessed by the growing crystal. And the stuff must be free to reach the crystal's surface.

In the growing of a crystal of alum in solution, aluminum sulfate and potassium sulfate diffuse through the water; and when they reach the surface of the crystal, they join with each other and with some of the water. They adopt positions on the surface that are forced on them by the kind of orderliness confronting them. Settling into those positions, they extend the orderliness outward, and thus the crystal grows.

Of course, nourishment cannot reach the parts of the surface resting on the bottom of the container, or butted against the sides, or in contact with the surfaces of other crystals growing in the same solution. This is why it is rare to see large single crystals, and especially rare to find them with many natural faces exposed and well developed.

Usually, both in nature and in man's manufacturing processes, a great many crystals of the same material start growing at about the same time in many different places. They grow until something gets in their way, or until they get in one another's way, and then they stop. Since they start with no knowledge of one another, they all have different orientations, and when they meet they cannot join to form a single big crystal. The result is a *polycrystalline mass*. Its component crystals all have the same kind of orderliness, but they all have different directions of that orderliness, as Figure 5 shows.

It is especially easy to see why metals, in the forms in which we use them, are nearly all polycrystalline masses. Almost always they are solidified by fairly fast cooling of the molten metal. When the molten metal cools to its melting point, which is the same temperature as its solidification point, innumerable little crystals sud-

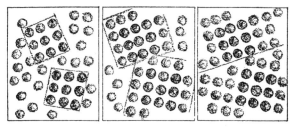

Fig. 5. TWO CRYSTALS GROWING *independently from a molten material do not join to form a single crystal. They will have the same pattern of orderliness, but one ordered group of atoms will be turned in a different direction from the other.*

denly form, and grow quite rapidly until they touch one another in all directions and no metal is left molten. Plate 6 shows the resulting arrangement of crystals in an ordinary brass doorknob.

You can watch the sort of thing that goes on when a molten material solidifies by experimenting with substances more convenient to handle than metals. A particularly suitable substance is "salol" (which the chemist calls phenyl salicylate), obtainable at a drugstore. Its melting point (43° centigrade) is low enough for you to melt it by putting a glass bottle of it in hot water. When it cools through its melting point again, the crystals grow slowly enough for you to watch the process at leisure through the sides of the bottle, as in Plate 7.

If you follow the directions in the caption to Plate 7, you may be interested to contrast the growth of salol crystals from the molten state with the growth of the same material from a solution of it in alcohol. As you might expect, the growth from the molten material is more rapid. There the molecules which will form the crystal are already closely packed together, and need only to lose enough of their random activity to settle into orderliness. In solution the molecules are diluted with molecules of alcohol, and they must travel to the surfaces of the growing crystal, pushing their way around and past those foreign molecules.

If you try removing crystals from the solution, and also from the molten material while it is still solidifying, you will see that a high speed of crystallization can be a disadvantage when you want to prepare a good single crystal. Molten material, adhering to a crystal when you remove it, solidifies before you can wipe it off, giving you a shapeless blob of solid. The relatively slower rate of growth from solution leaves you time to wipe off the adhering liquid and preserve the flat natural faces of the crystal.

It is interesting to make an estimate of how rapidly molecules must get themselves ordered at the surface of a growing crystal. You find that, even when the growth rate of material on a face is no faster than one or two millimeters per day, about a hundred layers of molecules must be laid down per second on the surface. And if the crystal is to be truly perfect, all these molecules must be laid down in the right sort of orderliness.

Imagining the atomic hustle and bustle this implies, you will not be surprised that crystals are seldom perfect. Even crystals that seem perfect to the eye and under the microscope usually have imperfections on an atomic scale. The defects to which crystals are subject are of many kinds, some crude, some subtle. They manifest their presence in various ways, most of them too complicated to discuss in this book. But you will understand solids better if you learn a little about some of the simplest sorts of imperfections in crystals.

Disturbances of Order

The most obvious way in which the perfection of order in a crystal can be disturbed is by impurities. Of course, crystals can acquire impurities by growing around foreign particles and including them. But there is a more interesting way in which a crystal can become impure, with inclusions not detectable under the microscope.

Often the atoms in the crystal are arranged in the right order without interruption, but different sorts of atoms occupy the sites that would be reserved for one sort alone in a perfect crystal. In other words, "impurity atoms" have invaded the structure; the host and the impurity form "mixed crystals." When this happens, you can be almost certain that crystals of the two substances —host and impurity—taken separately have one and the same sort of order, and that their molecules have nearly the same size and shape, so that they can replace each other comfortably.

A beautiful example of the formation of mixed crystals is provided by alum, the double compound in which potassium sulfate and aluminum sulfate combine in one-to-one proportions in the crystal. A similar double compound, chrome alum, forms if you combine potassium sulfate with chromium sulfate instead of with aluminum sulfate. The chromium sulfate plays exactly the same role in this compound that the aluminum sulfate plays in ordinary alum; the chromium atoms in the one occupy the orderly positions which the aluminum atoms occupy in the other. Indeed, the crystals of the two different alums look remarkably similar, except that ordinary alum is colorless and chrome alum is deep purple.

A solution containing both of these alums will deposit mixed crystals of the two. If the solution contains mostly ordinary alum, and only a little chrome alum, so will the crystals, and their color will be a pale purple. The

Fig. 6. A MIXED CRYSTAL *can be compared to an orderly design in which one motif has a different color here and there.*

color of the deposited crystals deepens if you increase the ratio of chromium to aluminum in the solution.

You can visualize what goes on as these mixed crystals form by extending the picture of what goes on when the pure alum crystals form. Potassium sulfate, aluminum sulfate, and chromium sulfate diffuse to the surface of a growing crystal, and there each molecule of potassium sulfate combines with either a molecule of aluminum sulfate or a molecule of chromium sulfate, whichever is handier. The crystal accepts chromium or aluminum impartially.

The mixed crystal therefore contains the same number of potassium atoms as the sum of the aluminum and chromium atoms lumped together. Some of the sites normally occupied by aluminum will be occupied at random by chromium. If you think of the chromium and aluminum as alike, the crystal is perfectly orderly. But if you distinguish chromium from aluminum, the crystal has some disorder: the disorderliness of the chromium-aluminum replacements. In contrast to the orderly pattern of Figure 4D, the mixed crystal is analogous to the pattern shown in Figure 6.

The partial replacement of aluminum by chromium, and sometimes by iron, is an important feature of many crystals found in nature. Rubies are crystals of aluminum oxide which acquire their prized color by including a little chromium in place of aluminum, and today rubies are artificially produced by growing crystals from a molten mixture of aluminum oxide and chromium oxide.

Sometimes one substance can partly replace another in a crystal, as suggested in Figure 7, even when the two

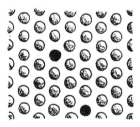

Fig. 7. A "SUBSTITUTIONAL IMPURITY" joins its host by substituting its atoms for some of its host's in the orderly arrangement.

pure substances do not have the same atomic arrangement. But in these the impurity can be included only in small proportion. The host can force compliance on only a few guests and rejects the rest; you cannot make crystals with *any* proportion of the two ingredients, as you can make alums with any proportion of aluminum and chromium.

Recently it has been found that in crystals of germanium some germanium atoms can be replaced by arsenic atoms, or by gallium atoms, even though crystals of pure arsenic and of pure gallium have atomic arrangements quite different from crystals of germanium. These "doped" germanium crystals, and similarly doped silicon crystals, are the essential ingredients of "transistors," which are now replacing vacuum tubes in many electrical uses.

Atoms like those of arsenic substituted for germanium in a germanium crystal are often called "substitutional impurities" to distinguish the way they are included in the crystal. Atoms of different elements differ in size, and if a crystal grows in the presence of small enough foreign atoms, some of those atoms may lodge in little open spaces left in the orderly arrangement of the atoms of their host. Fitting into interstices—open spaces—in the

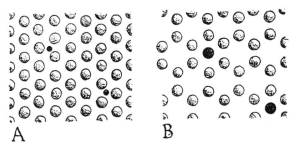

A   B

Fig. 8. An "interstitial impurity" *joins its host by placing its atoms in interstices in the arrangement of the host's atoms. Depending on the nature of that arrangement, the host may be able to accommodate only very small atoms (A) or quite large atoms (B).*

otherwise unaffected array of the host's atoms, these guest atoms are called "interstitial impurities."

How small is "small enough"? That will depend on the arrangement of atoms in the host. In the arrangement in Figure 8A "small enough" is very small indeed; in the arrangement in Figure 8B the interstitial atom could be as large as an atom of the host. The arrangement of atoms in germanium, for example, provides an open enough structure to accommodate hydrogen as an interstitial impurity.

Even when no impurity whatever is present, a crystal can acquire defects. You can think of one way in which it can do so by imagining a substitutional impurity that is nothing at all! Some of the atoms that must be present to give the crystal perfect order are simply omitted, leaving some of the atomic sites vacant, as shown in Figure 9.

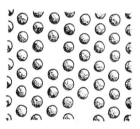

Fig. 9. VACANCIES *occur when some of the atomic sites in the regular arrangement are not occupied.*

If you grow crystals yourself, following the recipes we give, it will not surprise you that such vacancies can occur. Often, especially if you grow the crystals too fast, you will get crystals that appear milky. A microscope would show you that these "veils" are made of many little open spaces, filled with the solution from which the crystal grows. If you grow it slowly enough, the crystal will appear clear, but you can imagine that you might merely have reduced the size of the little holes enough so that they can no longer scatter the light to give the crystal a milky appearance. Supposing, finally, that you reduced these holes to atomic size, you see that they could no longer include any water and would represent vacancies. And remembering the atomic hustle which goes on at the growing surface, you will not be surprised then that before an atomic site can receive its appropriate atom, a new layer of atoms may clamp a lid over the site and leave it irrevocably empty.

The hustle of growth is not the only cause of vacancies. New vacancies can be produced in a finished crystal of some substances. Common salt—sodium chloride —is such a one. If you heat a crystal of sodium chloride quite hot—not far below its melting point—in an atmosphere consisting of sodium vapor, sodium atoms will slowly diffuse into the crystal. They add themselves to the inside, making the whole crystal slightly larger, and taking up the same sorts of orderly positions as the sodium atoms already there. But since no additional chlorine arrives to fill up the new chlorine sites created in this process, the newly created chlorine sites are "vacancies."

Thinking abstractly of all the kinds of disorder discussed so far, you could describe them as *points* of disorder. Interstitial impurities, substitutional impurities, and vacancies disturb the orderliness of the crystal at little isolated points dotted through it. Now let us turn to some kinds of defects that we can describe as *planes* of disorder.

In Figure 5, you have already been introduced to one of the most frequent and most important of these kinds of defects in a crystal: the "grain boundary" where two adjacent crystals join together in a solid. You may protest that this place is not a defect in either crystal; it is just the place where each crystal comes to an end. Indeed, if you prefer, you can look at this place as a defect in the polycrystalline solid rather than as a defect in any of its component crystals.

But however you prefer to think of the matter, notice a problem which the atoms face at the boundary between two crystals. They have a hard time deciding which crystal they belong to. A few layers of atoms along such a boundary will often have a disordered arrangement: the allegiance of each atom to one crystal or the other would be hard to identify, as Figure 10 shows.

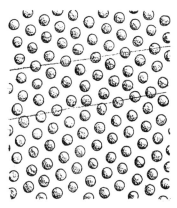

Fig. 10. A GRAIN BOUNDARY *between two differently oriented crystals of one material can be regarded as a kind of crystalline defect, in which a few layers of atoms are uncertain to which crystal they belong.*

In fact the atoms at *any* boundary of a crystal must have a slightly different arrangement from the atoms inside it. You can see this by contrasting the atomic situation inside with the situation at the surface of a crystal out in the air. The attractive forces between the atoms, which hold them together in solid form, are pulling on any one atom in all directions when it is inside. But an atom at the surface is attracted only in the inward directions. Hence, you can expect that in most crystals a few layers of atoms near the surface will be slightly more compressed inward than the bulk of the crystal.

After finding points of disorder and planes of disorder, you will surely ask, "Are there also defects which can be described as *lines* of disorder?" It is only rather recently that such defects have been found; they are called "dislocations." A very active study of them is showing that they are in many ways the most important defects that a crystal has.

The presence of dislocations in the component crystals of metals explains why the metals can be bent and pounded and rolled and drawn into different shapes without breaking. Dislocations even explain why some crystals grow when previous theories said that those crystals should not grow at all. If you want to pursue that fascinating subject, several articles and a book mentioned in the bibliography will help you.

At this stage of the discussion the mathematically minded may say, "A point has no dimension, a line has one dimension, a plane has two dimensions, and I have found types of disorder which correspond to all these abstractions. Nature seldom stops short of all possibilities: there ought to be types of disorder in *three* dimensions."

What would a "volume of disorder" be? It would be disorder everywhere in the crystal. In other words, it would be noncrystalline matter. You have already noticed that gases and liquids are disorderly throughout, and that solids are crystalline. But there is a kind of "solid" that is not crystalline—*glass*.

Perhaps "glasses" would be a better word than "glass," for there are more kinds of glass than most persons think. To see how they come about, visualize what happens as you cool a molten material made of atoms which are tightly connected together in groups to form molecules. The molecules are less agitated as the temperature goes down, and finally try to settle into orderly positions. But if the molecules are large and have an irregular shape, they may have difficulty getting into those positions. Each molecule may have to turn around, in order to jockey itself into the right place, and in this effort it may get so entangled with other molecules that it never succeeds.

Even in liquids you can see the effects of these difficulties. As you cool almost any liquid, it becomes more "viscous": it pours more slowly, and solid bodies do not fall through it as rapidly. Maple syrup fresh from the refrigerator pours very slowly indeed, and in winter you change to a "lighter" lubricating oil in your automobile because the summer oil would flow too sluggishly at the lower temperature. The increasing viscosity of liquids as their temperature goes down is a measure of the increasing difficulty with which the molecules move past one another.

In short, two factors control how well a molecule can accomplish its desires. It needs freedom and time—freedom to move, and time in which to move. The bigger the molecule and the more irregular its shape, the more of *both* freedom and time it needs. Ordinary glass is a material needing more freedom and time than anybody usually gives it. It will sit at ordinary temperatures for a lifetime, a mass of disordered, tangled molecules, quivering and unable to move enough to crystallize.

But glass manufacturers know that ordinary glass wants to crystallize nevertheless. The melting pots in which glass is made will sometimes acquire crystals on their edges after long periods of use. There, over long times, the glass has been so hot, and the molecules therefore so much freer and more agitated, that they have slowly found their way into orderly arrangements.

This gives you a hint for making glasses out of some

materials that ordinarily crystallize very easily. Sudden cooling may remove the molecular agitation in too short a time to permit crystallization. Sugar, for example, is made of big irregular molecules, which have an orderly arrangement in the crystals in your sugar bowl. If you pour molten sugar on a cold surface, it forms a glass. It is especially easy to make a glass by chilling molten boric acid.

Despite the rigidity of glasses, the physicist still wishes to reserve the word "solid" for crystalline solids. He may say, "Glass is rigid but not solid." Or he may call glass a "noncrystalline solid." He used to call glass a "super-cooled liquid"; and he has even been known to assign glass to a fourth state of matter: the "glassy state."

Clearly the physicist's difficulty in assigning glass to one of the three familiar states of matter comes from the fact that glass is like a solid in its rigidity and like a liquid in its disorderliness. Its molecules cannot move past one another easily, but they are vibrating about disorderly rather than orderly positions, like the positions shown in Figure 11.

In any case, "solid" or not, a glass is not crystalline. The cut-glass punch bowl, which is "crystal" to the shop-

Fig. 11. THE ATOMIC ARRANGEMENT IN A GLASS *has no order extending a long distance, but there is usually some short-range orderliness. In this example each black atom has three white atoms around it, at a fixed distance and at the corners of an equilateral triangle.*

keeper, is not crystal to the physicist. "Crystal gazers," who used to look into the future through spheres polished out of large single crystals of quartz, often look today through spheres of glass, because they are cheaper. It would be interesting to know whether the future seems as clear through a disorderly material as through an orderly one.

The Genesis of Minerals

If someone suddenly said to you, "Find a crystal, and be quick about it," you would probably forget the sugar in the bowl, the salt in the shaker, and leap out of doors to hunt for a glittering rock. The minerals of which rocks are made furnish the most familiar examples of crystals; everyone recognizes quartz, gems, and most semiprecious stones as crystalline. But it is less familiar that the entire solid crust of the earth is crystalline, with little exception. Indeed, most of the crust will show this to a sharp eye, aided here and there by a small magnifying glass.

The geologist makes a useful distinction between a "rock" and a "mineral." To him a mineral is a single substance in crystalline form. A rock is made of minerals—perhaps many minerals, perhaps only one. Granite, the rock shown in Plate 8, is made of three minerals: quartz, feldspar, and mica. Marble is a rock made of the single mineral, calcite.

Just how the earth arrived at the form in which we find it is a question still far from settled. Nobody was there when the business began, and we must now infer what might have happened from what lies before us. Consequently, we can speak with greater assurance about geological processes that have occurred relatively recently than about processes of the more distant past.

We find the continents and oceans supported by a

deeper-lying layer of dark rock, about six hundred miles thick, which many geologists believe floats in turn on a core of molten iron and nickel. It is easy to imagine that the dark layer, and the material of the continents, may also have been largely molten at one time, floating to the surface of the metal much as oil floats on water. In any event, some of the rock is molten today, and spurts to the surface of the earth in volcanic eruptions.

Solid or molten, the rock consists mostly of chemical compounds of silicon, aluminum, and oxygen, and smaller amounts of compounds of the other chemical elements. The deep-lying molten brew of silicates—the "magma"—percolates upward through cracks and crannies, sometimes dissolving rock already formed. When it is cool enough, some substances crystallize out, and if it cools slowly the crystals will be quite large. Losing some of its components in this way, the molten mass changes character: there are more substances with low melting points left. The mass of magma, still percolating up, may transport with it some crystals already formed. Finally what is left is mainly water, melting at a low temperature, carrying silica, silicates, and any salts that remain dissolved in it. As the magma nears the surface, cools and evaporates, these salts finally crystallize out also.

A great variety of rocks—the "igneous rocks"—are formed in this process. You can recognize them by examining a freshly broken surface: the grains of crystal within the rock are angular and sharp, interfering with one another. Some of the crystals are quite large, many smaller crystals are still visible to the unaided eye, and most of the rest appear under a magnifying glass. Granite, for example, usually displays clearly recognizable crystals of its component minerals: quartz, feldspar, and mica.

But many rocks formed in this way do not survive:

things happen to change them. One sort of change, mechanical wear, produces the great class of "sedimentary rocks." In dry climates the alternations of heat by day and cold by night crack the surface of the rock enough for wind to pick up the fragments and batter the rock with them still further. In humid climates the expansion of water freezing in cracks does the initial damage. Streams of water carry the fragments away, tumbling and rubbing them against one another until they are even smaller than the original crystals.

Fragments of any single substance, carried the same distance, will have about the same size. Since they also have the same density, they will settle to the bottom of the stream or the ocean all together as the current slows to a pace that no longer supports them. Thus there is a tendency for each substance carried by the water to settle in a bed of fragments of its own sort. Seashore sand is a good example of this—thousands of similar-sized pieces of quartz deposited together when the current had slowed just enough so that they dropped out. Although their outward shapes are not the shape of a quartz crystal, each is nevertheless a crystal.

Often a bed of rounded, broken, sedimentary sand will be further changed by water with dissolved minerals passing slowly through it. The minerals crystallize out of the water onto and between the granules of sand, cementing the loose aggregate together into rocks such as the sandstone shown in Plate 9. When very finely divided sediments build up into deep beds at the mouths of rivers, the weight of the whole mass may hold the particles together. The result will be shale, whose crystalline particles are very small indeed. Usually the grains of crystal in a sedimentary rock, unlike those in an igneous rock, will not be angular and sharp. They will reveal in their roundness the fact that they once tumbled about independently.

Both igneous and sedimentary rock can change in another way, to form the "metamorphic rocks." In metamorphism the crystalline minerals already present are remade into different crystal forms. Sometimes great pressures assist in reordering the atoms. Without the addition of any new substances, limestone becomes marble, and shale becomes slate. When silicious juices also play a part, "mineralization" of all kinds occurs.

Volcanic glass and opal are examples of the very few noncrystalline parts of the earth's crust. All metals, naturally formed and man-made, are crystalline, and occasionally large metal crystals appear in nature. To be sure, minerals usually lack the flat faces and geometric forms of crystals freely grown. But a few perfect, large crystals are found, such as quartz, calcite, and topaz. And small but well-developed crystals grow fairly often from mineral-rich water solutions moving through pockets, crannies, and caves deep inside the ground. Indeed, whenever the crystals can form slowly and have room to grow, they show their wonderfully organized structure in the development of their faces. Anyone who studies such snow crystals as that shown in Plate 10, or who visits the collection of minerals in a museum of natural history, cannot fail to be impressed by the symmetry adopted by these orderly organizations of atoms.

Lack of time and lack of freedom—these are the only lacks that have stood in the way of separating each of the solid materials of the world in the form of a single enormous crystal of its substance. Ice is an exception, of course, because the temperatures on earth happen to fluctuate up and down through the freezing point of water. The orderliness, laboriously constructed when the temperature falls below freezing, is bombarded into disordered rubble when the temperature rises again.

Ions and Salts

To many "salt" is a flavoring for food, and "salts" a perfume for the bath. To the chemist "salt," or "common salt," is only the most familiar member of a large class of substances which he calls "salts" and which have certain characteristics in common, distinguishing them from other substances.

Not all salts are soluble in water, but most of them are. It is for this reason that the "magma," cooling as it percolates upward toward the earth's surface, finally consists mostly of water containing dissolved salts. Sea water contains many salts, but principally common salt. Great Salt Lake, in Utah, contains salt which has been concentrated by the evaporation of the water. And there are places in the world where the evaporation of salty water in the past has left crystalline deposits of salt. Common salt is consequently a familiar mineral to the geologist; he calls it "halite" or "rock salt."

Another salt even more familiar to the geologist is calcite. It is not very soluble in water, but it dissolves a little more readily if the water also dissolves some carbon dioxide gas. Calcite is the salt composing chalk, limestone, marble, and the remarkable columns of rock in some caves, called stalactites and stalagmites. It is the principal offender in most "hard water"; the dissolved salt reacts chemically with the dissolving soap, destroying the suds. And it is a major ingredient in "boiler scale," the deposit which slowly accumulates on the inside of a steam boiler. The higher temperature inside the boiler drives out of the water some of the carbon dioxide which the water dissolved at a lower temperature; then the water will hold less calcite, and some of it comes out of solution.

In the home or the laboratory, just as in nature, most

of the substances that readily form crystals at room temperature belong to the class of salts. For that reason, the rest of this book will talk almost entirely about substances of that class. The best way to understand the special character of salts is to examine first a special character which their component atoms acquire. For definiteness, turn again to common salt. Already in this book it has proved to be a useful example of crystalline behavior, and now you can take it as an example illustrating the nature of salts in general.

A chemical analysis of common salt shows that it is made of sodium chloride—one atom of sodium for each atom of chlorine. You might expect that it would be made of *molecules* of sodium chloride, each molecule consisting of an atom of sodium and an atom of chlorine tightly joined together. You get the first suspicion that you are wrong when you dissolve the salt in water.

Common salt, like most salts, dissolves readily in water, and you find that the solution conducts electricity quite well—very much better than pure water. If you put two electrodes in the solution and connect them to a battery to pass a direct current through the solution, as shown in Figure 12, chlorine gas bubbles off at one electrode, hydrogen gas bubbles off at the other electrode, and the solution around the second electrode becomes more and more alkaline with sodium hydroxide.

So the chlorine and the sodium have been separated somehow. Has the electric current split the molecules? Has the water split the molecules? The best explanation is that there were no molecules in the first place—that the sodium atoms and the chlorine atoms were already separate. But why would the electric current pull one sort of atoms to one electrode and the other sort to the other electrode? It would do this only if the atoms were electrically charged, sodium with one sign, chlorine with the other.

If you do not see the reason for that conclusion at once, pause for a moment in your pursuit of sodium chloride, and call to mind what you know about electricity. You will remember that a current of electricity is a motion of electric charges, just as the flow of a river

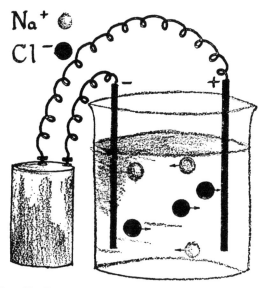

Fig. 12. THE OPPOSITELY CHARGED IONS *of sodium chloride in solution drift in opposite directions under the forces applied by an electric battery, and so carry an electric current through the solution.*

is a current of atoms tied together in water molecules. Atoms come in great variety, but electric charges come in only two kinds, "positive" charges and "negative" charges.

These electric charges are not disembodied things. They are properties of matter, and one of the greatest triumphs of physics near the beginning of this century was a leap in understanding how these charges are embodied in matter. The nucleus of an atom has positive charges, and electrons have a negative charge. The nuclei of different kinds of atoms have different amounts of positive charge, but all electrons are alike: each has the same amount of negative charge as any other. Mostly the electrons circulate around the nuclei in the atoms.

Under some conditions the electrons can separate themselves from the nuclei and wander around independently. In a television tube, for example, electrons come out of the hot filament in the tube and wander in the empty space. In a metal some electrons are so loosely bound to their parent atoms that they can wander through the metal. Metals conduct electricity well for that reason; a current of electricity through a metal is a flow of its loosely bound electrons, which thread their way among the atoms.

The reason for calling the two kinds of electric charges "positive" and "negative" is that the two behave in opposite ways, and indeed can cancel each other out. A body having equal amounts of positive and negative charge has no net charge at all, just as the addition of plus nine and minus nine gives you not eighteen but zero. And a flow of negative charge in one direction gives you the same electric current as an equal flow of positive charge in the opposite direction. Even though the opposite charges are moving in opposite directions, the electric currents their flows produce will be in the same direction. But if the opposite charges move in the same direction, their flows will cancel electrically, and there will be no electric current—only a current of the matter which is carrying the equal and opposite charges.

Normally, therefore, a piece of matter is "electrically neutral"—with no electric charge—not because there is no charge in it, but because the positive and negative charges in it are exactly equal in total amount. Indeed, if you subdivide the matter into its ultimate atoms, you will often find that those atoms are electrically neutral. The positive charge on the nucleus of each atom is exactly balanced by the total negative charge of all the electrons circulating around it.

The fact that matter—even an isolated atom—usually is electrically neutral points to another important property of electric charges. Electrically charged objects exert forces on one another. Two negatively charged objects, for example, try to push each other apart; so also do two positively charged objects. But two objects bearing charges of opposite sign try to pull each other together. The rule is: Like charges repel, and unlike charges attract.

Hence, if an atom has "one too few" electrons, so that it has a positive charge, it will attract any electron nearby, and usually capture it. If an atom has "one too many" electrons, it will repel all other electrons, and indeed it is usually a sitting duck for the loss of its extra electron to some other atom having one too few.

And now, returning to the experiment with the salt solution and the battery, you see that sodium chloride, in solution at any rate, furnishes an exception to the general rule that atoms are electrically neutral. It consists not of molecules, nor of electrically neutral atoms, but of *electrically charged* atoms. By drifting through the liquid, they carry electric charge from one electrode to the other.

When atoms are electrically charged, they are called "ions." If you notice which electrode in the solution the ions move to, you can decide whether their charge is positive or negative. It turns out that the sodium ions are positively charged and the chloride ions are negatively charged. Clearly what has happened is the transfer of an electron from each sodium atom to a chlorine atom, leaving the sodium a positive ion and making the chlorine a negative ion.

Why do sodium and chlorine atoms behave in this exceptional way when they get together? You would expect each positive sodium ion to grab an electron from each negative chloride ion, and produce electrically neutral sodium and chlorine atoms. Chemists and physicists have learned some of the reasons for this unexpected behavior but the explanation of it is too long for this book.

Even when you have accepted the idea that sodium chloride splits into ions in solution—"dissociates," as chemists say—you may be surprised that there are not molecules of sodium chloride in solid salt. But you can see some of the evidence that there are no molecules by looking at the arrangement of atoms in a crystal of sodium chloride.

The positions of those atoms are arranged in the pat-

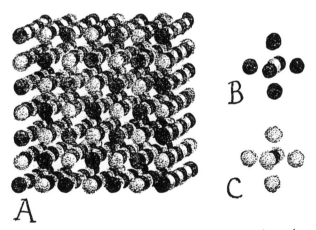

Fig. 13. A SODIUM CHLORIDE CRYSTAL *consists of an orderly arrangement* (A) *of sodium and chloride ions in which molecules of sodium chloride cannot be distinguished. Each sodium ion is surrounded by six chloride ions* (B), *and each chloride ion by six sodium ions* (C).

tern shown in Figure 13. Each sodium is equally surrounded by six chlorines and each chlorine by six sodiums. A sodium atom owes no more allegiance to one of its six neighbors than to another. If there were molecules of sodium chloride in the solid, you would expect to see some evidence in the atomic arrangement that sodium atoms were paired off with chlorine atoms.

And now if you suppose that the sodium and chlorine atoms are ions even in the solid, you can see why the solid holds together. Recall the attraction between opposite charges and the repulsion between like charges, and notice in Figure 13 that each positive ion has collected negative ions around it, and vice versa. Moreover, each positive ion is keeping the other positive ions as far away from it as possible, and each negative ion is doing the same thing to the other negative ions. In other words, the electrical attraction between oppositely charged ions is the binding force in sodium chloride; and that attraction, combined with the electrical repulsion between ions of the same kind, largely determines the regular arrangement which the ions take in the crystal.

The chemist calls those substances salts which he can obtain by mixing an acid and a base. In the case of sodium chloride (NaCl), the acid would be hydrochloric acid (HCl), and the base would be sodium hydroxide (NaOH). The product, sodium chloride, dissociates into ions in solution and even in its solid form, and all other salts show much the same behavior.

The physicist calls the resulting solids ionic crystals. Indeed, he calls any crystal an ionic crystal in which he has reason to believe that the atoms bear net electric charges, as do the atoms in salts. But the physicist's ionic crystals and the chemist's salts are not altogether the same. Magnesium oxide forms ionic crystals, but the chemist would not call any oxide a salt, because it does not contain the residue of an acid.

More often than not, either the basic residue in a salt, or the acid residue in it, is made of more than one atom. Ammonium chloride (NH_4Cl), for example, is a salt not unlike sodium chloride, but the basic part of it—the "ammonium" (NH_4) part—is made of five atoms: an atom of nitrogen to which four atoms of hydrogen are tightly bonded. Conversely, in calcium carbonate ($CaCO_3$), the substance whose solid form is calcite, the acid "carbonate" part is made of one atom of carbon and three of oxygen.

In these substances the ions are like little molecules. But they are little molecules with a net charge. It would be idle to ask which atom in the little molecule has gained or lost the extra electron. The ammonium ion is simply an assembly of five nuclei and a lot of electrons. Some of the electrons circulate around individual nuclei, others around pairs of nuclei. And there is one electron less than the number required to make the little molecule electrically neutral: it has a net positive charge.

In calcite each little negatively charged molecule containing carbon and oxygen has gained two extra electrons from a calcium atom, not just one. The arrangement of atoms in each carbonate ion places the carbon at the center of an equilateral triangle of three oxygens. The ion is "flat," though of course it has thickness.

In alum—the double compound of potassium sulfate and aluminum sulfate—the sulfate ion is again complex, consisting of a sulfur atom and four oxygen atoms. Like the carbonate ion, the sulfate ion has two extra electrons. A pair of sulfate ions gains one of its four electrons from a potassium atom and three from an aluminum atom, leaving a potassium ion with charge plus one, and an aluminum ion with the unusually large charge of plus three.

But alum crystals contain more than these ions. They

also contain water! The water molecules are tightly bound into the crystal, and have their own orderly part to play in it. Indeed, many crystals, especially crystals of salts, contain orderly water in their composition, and are often called "salt hydrates." In the next chapter you will learn how this comes about.

CHAPTER II

Solutions

Crystals grow under the most various conditions, some of them quite amazing. Snow flakes grow directly from moist air—in other words, from a gas. Large crystals of metallic bismuth grow from the molten metal—in other words, from the same substance in liquid form. Large tungsten crystals grow directly from polycrystalline tungsten when it is heated and stressed at the same time —in other words, from a solid. But the most familiar method of growing crystals is from a solution in a liquid —in other words, from a second substance which stays liquid or evaporates while the solid separates out.

A wide variety of interesting crystals form in this way in chemical compounds of the class called "salts," the substances we have described in the previous chapter. And almost all salts can be crystallized from the most readily available liquid, water. For this reason the recipes in a later chapter deal only with the crystallization of salts from their solutions in water. But there are many other kinds of solutions, and you will understand better the nature of salt solutions if you look first at solutions in general.

The Nature of Solutions

The word "solution" brings to most persons' minds a solid dissolved in a liquid; sugar dissolved in tea, or salt dissolved in broth. It brings these things to the minds of the chemist and the physicist too. But they will use the word rather more broadly than most laymen would.

For example, they think of gases also as dissolving in liquids, and perhaps you will agree with them when you remember how air bubbles come out of water when you heat it. The air in those bubbles had been in solution in the water. Carbon dioxide fizzes out of soft drinks when you release the pressure. Acetylene gas dissolves so abundantly in a liquid called acetone that it is usually shipped in tanks containing wood chips soaked in acetone. Indeed, the chemist and physicist think of a substance as dissolving in something else, to form a solution, whenever that substance is broken apart into its constituent molecules, and those molecules are dispersed as independent little units through the other material.

But they will use the word "solution" only when the dissolved substance is truly fragmented into particles no larger than molecules. When the substance is dispersed in larger fragments, they call the product a "mixture." Graphite grease is a mixture of very fine particles of carbon in grease. Cream is a mixture of very fine globules of butterfat in water. To be sure, the water in cream has other things, such as milk sugar, *dissolved* in it; but the fat globules are *mixed* in it, in the form of extremely small droplets.

Not only solids and gases but also *liquids* dissolve in liquids. If you pour a little alcohol into water, the alcohol dissolves in the water; and if you pour a little water into alcohol, the water dissolves in the alcohol.

The dissolving of alcohol in water calls attention to two more words often used in speaking of solutions. The water might be called the "solvent"—the substance that does the dissolving—and the alcohol might be called the "solute"—the substance that gets dissolved.

But you can see that these words are much more useful when the solute is something like a salt than when it is alcohol. With alcohol and water you get a single clear solution no matter what proportions of alcohol and water you use. Which is the solvent and which the solute? There is really no difference, and it would be a pity to make a distinction in words when there is no corresponding distinction in physical behavior. "Solvent" and "solute" are useful words when the solvent will dissolve only a limited amount of solute. The limited amount of solute a solvent will dissolve is called the "solubility" of the solute in the solvent.

There is an interesting difference between the solubilities of *gases* dissolved in liquids and those of *solids* dissolved in liquids. When you heat water, some of the dissolved air comes out of solution. But you may have noticed that when you heat water, you can dissolve more sugar in it. These are instances of a fairly general rule: Gases are usually less soluble in a liquid when it is hot than when it is cold, but the reverse is usually true of solids.

These opposite effects of heat on the two types of solute result from the same cause. Increasing the temperature of things tends to increase their disorder. The molecules of undissolved solid get more disordered by dissolving; the molecules of dissolved gas get more disordered by escaping from the solution and becoming a free gas.

It is easy to see how dissolving a solid disorders the molecules: the orderliness of their average positions in the crystalline pattern disappears, as Figure 14 shows.

In the case of a gas, the increase of disorder on escape from the solution is more nearly one of motion than of arrangement. In solution the molecules of gas have a disordered arrangement, but their motion is restricted

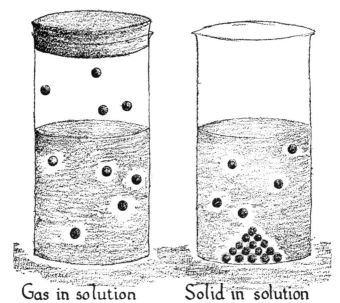

Gas in solution   Solid in solution

Fig. 14. MOLECULES IN SOLUTION *are caged by the solvent, which restricts their motions. Nevertheless dissolving a solid increases the disorder of its molecules, because the orderliness of crystalline arrangement disappears. On the other hand, dissolving a gas reduces the disorder of its molecules, because the liquid cages reduce their freedom of motion.*

because they are confined by the tightly packed molecules of the liquid surrounding them. You can think of each gas molecule as caged by the liquid molecules immediately around it. It can bat back and forth rapidly in the cage, while the cage moves about rather slowly through the liquid. When it escapes from its cage into a gas, it can roam with high velocity, almost at will, among its neighbors. Figure 14 suggests the contrast in the cir-

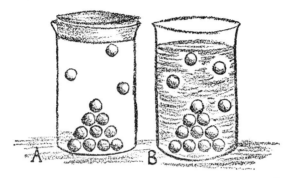

Fig. 15. Molecules of dissolved solid *are in somewhat similar circumstances when dissolved in a gas (A) and in a liquid (B), except that the liquid restricts their motions much more.*

cumstances of a molecule when it is dissolved in a liquid and when it is free.

You may ask, "Don't substances also dissolve in gases, and also dissolve in solids?" Clearly everything that can be vaporized dissolves in *every* gas. When it vaporizes into an environment containing a different gas —the air, say—then the vapor dissolves in that gas, unless the vapor and the gas react chemically and make a new substance out of both.

In some ways the solubility of a solid in a gas is like the solubility of a solid in a liquid, as Figure 15 suggests. When a solid vaporizes into a surrounding gas, its molecules are dispersed into a space thinly occupied by other molecules. When a solid dissolves in a liquid, its molecules are dispersed into a quite densely occupied space. But in both instances the ability of the substance to disperse increases with temperature. The most conspicuous difference is that the solubility in a liquid is much more selective. Common salt dissolves in water but very little in alcohol; salol dissolves in alcohol but very little in water.

The solubility of a substance in solids is even more selective than its solubility in liquids. In the last chapter you met an example of one substance dissolved in another in solid form: the "mixed crystals" of ordinary alum and chrome alum. The frequently used term "mixed crystal" is an unfortunate one because it suggests a mixture rather than a solution. "Solid solution" is a better term; the two alums form a true solution, for they are dispersed through one another molecularly, not in little chunks. Any amount of one alum will dissolve in any amount of the other, much as alcohol and water will dissolve each other in any proportion. In the case of substitutional impurities, discussed in the last chapter, the solubility of the substituent is limited, much as is the solubility of salt in water.

The two alums are soluble in each other because separately they have the same atomic arrangement, and their molecules have about the same size and shape. But simultaneous correspondence of arrangement, size, and shape is rare. You find it most frequently among the metals. Silver and gold, for example, have the same atomic arrangement; they can form "alloys" in any proportion, retaining that arrangement, with silver atoms and gold atoms occupying the same sorts of sites in the orderly pattern.

Pure copper and pure zinc have different atomic arrangements. A little copper will dissolve in zinc to form crystals of an alloy which retains the crystalline arrangement of zinc. Quite a lot of zinc will dissolve in copper, substituting for it in the copper arrangement and forming one of the several kinds of brass. When you make alloys with more zinc than the copper will dissolve and more copper than the zinc will dissolve, you get new atomic arrangements—other kinds of brass. This situation may remind you of ordinary alum, where potassium sulfate and aluminum sulfate combine to form an arrangement different from that of either ingredient. But in alum the ratio of potassium to aluminum must be exactly one to one; in the brasses the ratio of copper to zinc can vary somewhat.

Now compare how fast the dissolved molecules can run around in these different sorts of solutions. When molecules of solute are moving about in a liquid, they cannot go as far in as short a time as they can when they are moving about in a gas: the rate of "diffusion" is much slower. For example, the scent of a perfume, evaporating at one side of a room, reaches the other side fairly soon, even when the air seems still, by rapid diffusion of the molecules of perfume through the air. To observe the much slower rate of diffusion of molecules through a liquid, put a little chrome alum in the bottom of a test tube filled with water, and see how long it takes for the purple color to reach the top.

This does not prove that the molecules of solute are moving slowly in a liquid. It only shows that they are moving *around* slowly and take a relatively long time to cover any distance. Even in a solid the molecules are moving fast—vibrating rapidly and getting nowhere. In a liquid they are moving rapidly in their liquid cages, while those cages move slowly about through the liquid.

Since the motion of the molecules in a solid is mostly vibration, they very seldom change places and get anywhere, and usually you cannot make solid solutions by putting two solids in contact with each other. A piece of silver held against a piece of gold will not diffuse into the gold enough to form an appreciable amount of the alloy within a lifetime. You can make the alloy only by melting the two metals together. Soldering depends on dissolving a little of the solid metal in the molten solder to form an alloy at the boundary; when the solder freezes again, everything stays put.

Clearly, then, there are at least two factors that influence whether a substance will dissolve in something else. Its molecules must become more disordered by dissolving, and they must be free to move into the more disorderly arrangement. But surely these two factors are not the only ones. If a solid goes into solution in order to get more disordered, then, since it can get even more disordered by vaporizing into a gas, why do not solids always evaporate even more easily than they dissolve? Something more than just desire for disorder and freedom to reach it is going on.

The clue to the nature of that additional something comes from the fact that you cannot dissolve just any solid in just any liquid. Remember that forces of attraction bring the atoms or molecules together to form a solid. The random motions of the molecules, tending to separate and disorder them, wage a constant war with their attractive forces. Unless you amplify those motions by raising the temperature, you can expect the desire for disorder to win the war only under two conditions. You must either reduce the forces between the molecules or invoke new forces strong enough to pull them apart.

A solvent does one or the other, or a combination of both, as indicated in Figure 16. Some solvents weaken the binding forces between the molecules at the surface of the solid, enough for their agitation to knock them away from one another. Swimming out into such a solvent, the atoms or molecules find that their mutual attractions are even further reduced, and they wander independently in the liquid.

Other solvents apply forces to the molecules of the solid strong enough to tempt them away from the solid into the solvent. You could think of the molecules of the solvent as forming something like a chemical compound with the molecules of solute. But it is a sort of compound in which the two kinds of molecules are so loosely bound together that the composition of the com-

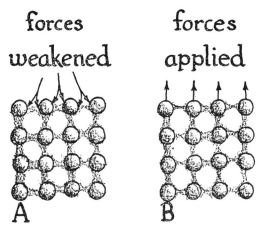

Fig. 16. TWO EFFECTS OF A SOLVENT ON A SOLID. *The solvent may dissolve it (A) by weakening the forces between its molecules, or (B) by applying forces which counteract the forces between its molecules, or both.*

pound is variable. The associations between the molecules of solute and of solvent keep forming and breaking and forming again.

Solutions of Salts in Water

In solutions of salts in water, a combination of both of these processes goes on: a weakening of the attractions between the constituents and a forcible removal of them from the dissolving solid. Recall from Chapter I that the ultimate constituents of salts are ions. The ions are atoms bearing electric charges, and electric forces between the ions hold them together as solids. Water can weaken the electric forces between the ions and also attract the ions away from the solid.

The great ability of water to reduce the electric forces between objects carrying electric charges is one of its most important physical properties. All liquids have this property to some extent, but water reduces the electric forces more than almost any other liquid does. This is the first part of the reason why water is an especially effective solvent for materials whose molecular constituents bear electric charges.

Contrast the resulting solubility of common salt with that of salol. Salol is made of electrically neutral molecules. From its low melting temperature (43° centigrade) you can guess that those molecules are held together less tightly than the ions of salt; they can separate and move past one another more easily. Nevertheless, salt, which melts at about 800° centigrade, dissolves much more abundantly in water than salol does. The water greatly reduces the forces between the charged ions which compose salt, but it reduces the forces between the uncharged molecules of salol very little.

The second part of the reason why water is a good solvent for most salts is that it tends to combine with their ions. A bare ion attracts a few water molecules, and then moves about in that clothing. Joined by its group of water molecules, each ion forms something a little like a single molecule of a chemical compound. This loosely bound molecule still carries the same total electric charge that the original ion carried, but now that charge is spread out over the whole molecule. And it is this charged molecule, not the bare ion, which moves around through the water.

Water tends, more strongly than other liquids, to unite with ions in this way, to "hydrate" them. Usually the easiest way for the ions to satisfy this hydrating tendency is to go into solution in the water. You can think of the water as trying to get around the ions at the surface of the solid, forcing its way between adjacent ions, and prying them loose. Then each clothed ion looks, from the outside, rather like water, as Figure 17 shows. The

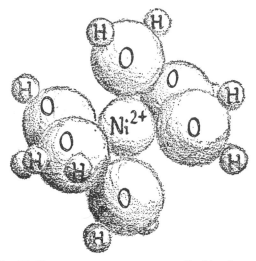

Fig. 17. THE HYDRATION OF AN ION. *Positive ions gather water molecules to them more readily than negative ions. Magnesium, zinc, and nickel are examples of positive ions; each gathers six water molecules around it when it is dissolved in water. The hydrated ion looks to the surrounding water rather like another bit of water.*

rest of the water accepts the clothed ion into solution, hardly recognizing the difference between it and another bit of water.

Sometimes, however, both of these actions are insufficient to dissolve a salt. For example, very little calcite dissolves in water. Sometimes the hydrating tendency of water is so strong that water molecules will join the ions of a solid and yet leave it solid, insoluble in the remaining water. This is what goes on in Portland cement, the familiar powdered cement used in making concrete, which sets into a hard insoluble block of solid when it is mixed with the right amount of water.

The behavior of another kind of cement, plaster of Paris, makes it especially clear that there must be still other factors that determine when a salt is soluble in water and when it is not. Plaster of Paris is calcium sulfate—calcium ions and sulfate ions. When you put it in water, it behaves somewhat as Portland cement does: it absorbs a definite amount of water to make a new insoluble solid. You might think that it behaves exceptionally because it contains some exceptional ion. But look at the following facts. Alum, which is a combination of sulfates, is soluble, so the special ion cannot be sulfate. Since calcite, which is calcium carbonate, is almost insoluble in water, you might be especially tempted to accuse the calcium ion of behaving exceptionally, until you notice that calcium chloride, the salt often used to reduce the dustiness of roads, is very soluble in water.

In fact, the problem of solubility will seem especially puzzling if you compare calcium sulfate, calcium chloride, silver sulfate, and silver chloride. Calcium chloride and silver sulfate are both very soluble in water. But if you keep exactly the same ions and switch them over, to form calcium sulfate and silver chloride, both of the resulting salts are almost completely insoluble in water.

Before your eyes, you will get dramatic evidence of this if you make two separate solutions, one of silver sulfate, the other of calcium chloride, and pour them together. A dense white precipitate of silver chloride and calcium sulfate will form immediately. The experiment shows you that the ions are moving quite independently in each solution, and rapidly find their way to form the least soluble combinations when the solutions come together.

The experiment also shows you that the solubility of a salt in water depends not only on what ions are present but also on what *combinations* of ions are present or can be formed. Even though the water climbs around the ions in calcium sulfate and hydrates them, the attractions between the clothed ions of calcium and of sulfate remain strong enough to hold them together in an insoluble solid. Solubility is a complicated matter indeed, and nobody understands it completely.

Along with alum, plaster of Paris is therefore another example of the salt hydrates mentioned at the end of the last chapter. By noticing that ions tend to be hydrated in water solutions, you can see why many other salts form salt hydrates. On arriving at the surface of a growing crystal, the ions are not bare; they are carrying a clothing which consists of a few water molecules. In order to form a crystal which contains no water, an "anhydrous" crystal, the ions must cast off that clothing.

This can be a hard job. The attraction of the ion for the clothing must be counteracted, and the torn bits of it—the water molecules—must then move out of the way. Often it is easier for the ions to crystallize into a salt hydrate, carrying some of or all their attached water molecules into the solid composition. When a salt can be crystallized in both an anhydrous and a hydrated form, the hydrate will usually grow more readily and more rapidly.

When the ions retain their watery clothing in the solid, the water molecules, like the ions, adopt an orderly arrangement in the crystals, as Figure 18 suggests. Then there is almost always a definite number of water molecules included per ion. If there were not, the arrangement could not be orderly. In alum there are twelve molecules of water per aluminum ion, and six of them immediately surround that ion.

Even when the number of water molecules is definite and the hydrated ions are combined into a crystalline solid—an orderly arrangement of ions and water molecules—they may nonetheless be able to combine in several ways to form several different solids. In each solid the number of water molecules per ion is definite; the solids differ in what that number is.

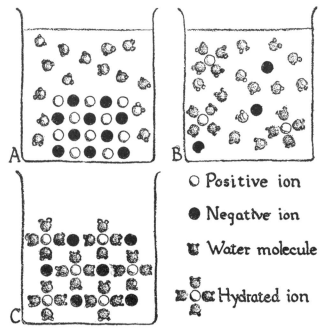

Fig. 18. THE FORMATION OF A SALT HYDRATE. *When the anhydrous salt comes in the presence of water (A), it dissolves, and the positive ions become hydrated (B). In crystallizing from solution again, the ions may remain hydrated, and then the water molecules play an orderly role in the crystalline salt hydrate (C).*

For instance, copper sulfate can form three different hydrates, with three, with five, and with seven molecules of water per copper ion. Recipes in a later chapter tell you how to prepare two hydrates of nickel sulfate containing, respectively, six and seven molecules of water per nickel ion. And these two hydrates have completely different orderly arrangements. There is no possibility of making solid solutions of one in the other and so getting a variable percentage of water in the crystal.

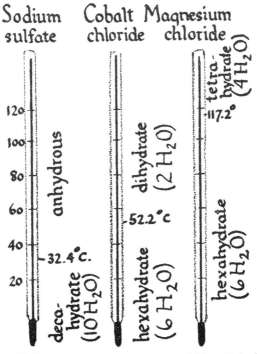

Fig. 19. THE TEMPERATURE RANGES *within which the different hydrates of a salt crystallize from water have definite limits. The hydrate with the most water crystallizes at the lowest temperature.*

When a salt can form several different hydrates, the hydrate containing the most water will crystallize at the lowest temperature. As the temperature is raised, the water molecules detach themselves more readily from the ions. In order to make crystals of the hydrates with the lowest water content, you must often operate at a temperature higher than normal. Examples of the ranges of temperature within which different salt hydrates crystallize from water are shown in Figure 19.

For the same reason, heating crystals of salt hydrates usually dehydrates them. With the escape of water, the crystals fall apart into minute crystals of a hydrate with less water, or even into the anhydrous salt. Unless the environment is very moist, the nickel sulfate crystals containing seven molecules of water per nickel ion—the "heptahydrate"—readily lose some of their water, even at room temperature. The nickel sulfate crystals with six molecules of water per nickel ion—the "hexahydrate" —are stable under ordinary conditions. But you can dehydrate the hexahydrate by heating it, to form a yellow powder consisting of anhydrous nickel sulfate. Plaster of Paris is a hydrate of calcium sulfate which becomes a higher hydrate when you add water at room temperature, and which can be dehydrated by heating to form plaster of Paris again.

Saturation and Supersaturation

Solutions present puzzling problems. The fact that some salts are soluble in water while others are not is one of them. Another is the puzzle of "supersaturation." When you grow crystals, you will make constant use of supersaturation, and you will need to know something about it.

First recall that the amount of any salt a given amount of water will dissolve is limited. When the water actually contains all that amount, it is said to be "saturated" with the salt; when it contains less, it is "unsaturated"; when it contains more, it is "supersaturated." You will object that, if the water can contain *more* than the limited amount, then that amount was not really the limited amount. But imagine the following experiment.

If you put solid alum into a given amount of water at room temperature, the water will dissolve the alum until it can dissolve no more. It may take a long time to reach this condition because, as the solution becomes more nearly saturated, it dissolves alum less and less rapidly. If you heat the solution after it has become saturated, it will begin dissolving more alum again because, with increasing temperature, the limited amount which the water can hold—the "solubility" of alum—increases also. If you finally pour the warm solution off any alum that remains undissolved, and then let the solution cool to room temperature again, you have every right to expect that the extra alum you dissolved at the higher temperature will come tumbling out of the solution in solid form. After all, lowering the temperature of the solution reduces the solubility of alum to the value it had before you heated the solution. And now the puzzling fact is this: maybe the extra alum does come tumbling out, and maybe it doesn't! If it doesn't, you call the solution "supersaturated."

Before examining this strange behavior further, look first at some facts that will reassure you that the "limited amount" still has a special status, that "solubility" has real meaning. If you put a pinch of solid alum in an unsaturated solution of alum, the solid will dissolve. If you put such a pinch of solid in a saturated solution, the solid will remain unaffected. And if you put such a pinch in a supersaturated solution, the solid will increase in amount. It will continue to increase until the solution is precisely saturated.

Perhaps "limiting amount" would be a better description of solubility than "limited amount." It is the content of salt which the solution *approaches as a limit,* either from the unsaturated side or the supersaturated side, so long as there is extra solid present to provide something that can either dissolve or grow. If the solution is very unsaturated, the salt will dissolve faster than if the solution is nearly saturated. Similarly, if the solution is very supersaturated, the salt will deposit faster.

In short, saturation represents a limiting condition which a solution approaches more and more slowly the closer it gets, from either the unsaturated or the supersaturated side. That slow approach makes it difficult to prepare a precisely saturated solution by adding or subtracting solid. To be certain that you have reached saturation, you must keep shaking the solution with excess solid for a long time, whether you are approaching saturation from "below"—by way of an unsaturated

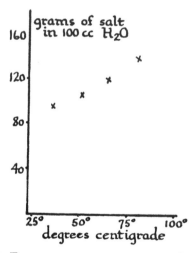

Fig. 20. THE MEASURED SOLUBILITIES *of a salt in water at various temperatures can be plotted as points.*

solution—or from "above"—by way of a supersaturated solution.

The difficulty of preparing a precisely saturated solution makes it difficult in turn to determine accurately the solubility of a salt. To measure a solubility, you must prepare a precisely saturated solution at a precisely known temperature. Then you can pour off or filter off some of the solution from the excess solid, weigh the solution, evaporate the water from it, weigh the solid deposited, and calculate the solubility.

After you have done this with solutions prepared at several different temperatures, you can plot the measured solubilities as points on a graph like Figure 20.

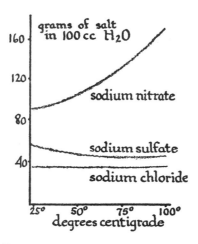

Fig. 21. THE SOLUBILITIES OF THREE SODIUM SALTS *behave differently with changing temperature. The curves are obtained by drawing lines through points such as those in Figure 20, which represent measured solubilities. The solubility of sodium nitrate increases with temperature; the solubility of sodium sulfate decreases with temperature; the solubility of sodium chloride is almost independent of temperature.*

In such a graph, the horizontal position of a point represents the temperature at which you saturated the solution, and the vertical position represents the limiting amount which you found for that temperature. Then you can connect the points by a smooth curve, on the assumption that the solubility, at temperatures between those which you actually used, does not jump around in an irregular way but changes smoothly.

In Figure 21 you see curves obtained in this way for the solubilities of three different salts in water. There you will notice that the solubility of sodium nitrate in water increases with increasing temperature. The solubilities of most salts behave in this way, responding to the disordering influence of heat. The solubility of sodium chloride in water is unusual in changing very little with temperature. Sodium sulfate is even more unusual: it is less soluble in hot water than in cold. If you want to pursue the interpretation of such diagrams further, the next chapter will help you.

Now look for a minute at the contrast between solutions on the opposite sides of saturation—the unsaturated and the supersaturated sides. An unsaturated solution is a relatively stable thing. If it is not in contact with any solid, and if it is not allowed to evaporate, it remains as it is for days on end, with no change of composition.

But a supersaturated solution is not in this stable condition. If you put a grain of the right solid in it, the same solid that has already been dissolved to form a supersaturated solution, more of that solid will form on the grain. The grain will not stop growing until the supersaturation is exhausted and the solution has reached precise saturation. In other words, you might consider an unsaturated solution as one that would like to become saturated, but the extra salt it needs is out of reach. A supersaturated solution would also like to become saturated, and can simply throw out its extra salt to achieve its ambition.

Then why does it not do so, promptly and completely? As a matter of fact, many supersaturated solutions do throw out their extra solids and reach saturation, especially when supersaturation is very high. But many will tolerate a small degree of supersaturation and give the appearance of stability. As soon as they come in contact with a little solid on which to throw their excess, they deposit more solid on it.

These apparently stable, slightly supersaturated solutions are very fastidious in their choice of a place to throw their excess. Usually they will respond only to a crystal of the substance with which they are supersaturated, or of a substance whose crystalline arrangement is like it. For example, a supersaturated solution of ordinary alum will deposit its excess on a crystal of alum or a crystal of chrome alum impartially. But it will not deposit on a crystal of potassium sulfate, even though potassium sulfate is one of the chemical ingredients of alum. If it is very supersaturated, it might deposit potassium sulfate on such a crystal, leaving the aluminum sulfate in solution.

One of the recipes you will come to will make it conspicuous that a supersaturated solution deposits in a fashion depending very critically on the kind of orderly atomic arrangement presented to it. Following that recipe, you let a supersaturated solution of nickel sulfate deposit its excess on a seed crystal of nickel sulfate hexahydrate (six water molecules), one of the two hydrates of nickel sulfate already described. Though the solution is even more highly supersaturated with respect to the heptahydrate (seven water molecules), no crystal of heptahydrate forms in the solution as long as no seed of heptahydrate is present.

After doing this experiment, you may wonder why a

supersaturated solution ever deposits crystals spontaneously. Apparently the dust in the air contains seeds with
an immense variety of crystalline atomic arrangements.
A nickel sulfate solution exposed to the air at ordinary
temperature will usually deposit crystals of the heptahydrate. The only way crystallization of the hexahydrate
can be carried out successfully at ordinary temperature is
by bottling up the solution while it is still unsaturated,
dissolving all seeds which the air carried in. Dust in the
air of a chemical laboratory is likely to be especially
rich in types of crystalline particles because materials of
great variety have been handled there.

Another reason why a supersaturated solution may
deposit solid is that it becomes more unstable as its
supersaturation increases. If it is very supersaturated, it
may deposit solid on almost any dust, especially on
spiky, sharp-pointed dust. A chemist who has prepared
an entirely new substance in a solution may have great
difficulty getting the substance to crystallize. But often
he can start crystallization by scratching the inside of
the flask containing the solution with a glass rod.

You may ask, "Why doesn't he just let a drop of the
solution evaporate to dryness and get his seeds that
way?" But what is dryness? The drop evaporates to a
smear, still perhaps containing some solvent, and so
viscous that the dissolved molecules no longer have freedom enough to arrange themselves in crystalline order.
Chemists working with compounds allied to sugar frequently have this distressing experience.

About ten years ago a company was operating a factory which grew large single crystals of ethylene diamine
tartrate from solution in water. From this plant it
shipped the crystals many miles to another which cut
and polished them for industrial use. A year after the
factory opened, the crystals in the growing tanks began
to grow badly; crystals of something else adhered to

them as shown in Plate 11—something which grew even more rapidly. The affliction soon spread to the other factory: the cut and polished crystals acquired the malady on their surfaces.

Enough of the unwanted material was collected to make a supersaturated solution of it. Since crystals of *both* materials—the unwanted and the wanted—would grow in that solution, the unwanted substance must contain the desired substance. And since crystals of both would grow in a pure solution made from the desired crystals, the unwanted crystals could not be the result of an impurity which had crept into the solution during the manufacturing process.

The wanted material was *anhydrous* ethylene diamine tartrate, and the unwanted material turned out to be the *monohydrate* of that substance. During three years of research and development, and another year of manufacture, no seed of the monohydrate had formed. After that they seemed to be everywhere. You can imagine, if you like, that in some other world nickel sulfate hexahydrate is well known, and the heptahydrate has not yet appeared. Perhaps in our own world many other possible solid species are still unknown, not because their ingredients are lacking, but simply because suitable seeds have not put in an appearance.

This curious behavior of supersaturated solutions, this illusion of stability, has a counterpart in molten materials. Chapter I has pointed out that glasses are molten materials cooled below their freezing points without crystallizing. Fog sometimes cools below the freezing point of water, and the droplets freeze when the windshield of a car strikes them. Mechanical shock may start crystallization in a supersaturated solution also. The chemist who has scratched the side of his flask with a rod will also shake the flask vigorously. So, in order to grow a large single crystal on a single seed, without allowing the solution to deposit more seeds spontaneously, keep the solution quiet.

What underlies this ability of melts to supercool, of solutions to supersaturate? Perhaps it is the difficulty of attaining orderliness. Without doubt, small groups of molecules moving about in the liquid occasionally fall into order, but this orderliness disappears before more molecules can join them. The infinitesimal crystal, present for an instant, melts or dissolves again.

In contrast, there is seldom any difficulty in attaining disorder; the difficulty of dissolving one solid in another is one of the few exceptions. You can often supercool a molten substance, but you can never "superheat" a solid —heat it above its melting point without melting it. You cannot put a solid in an unsaturated solution without dissolving it. And following any of the recipes we give, you will be impressed with the difference in the rates of crystallization and dissolution of a solid. It dissolves gleefully; it forms again, even when a seed is present, only laboriously.

CHAPTER III

Solubility Diagrams

The scientist is not the only person who systematically draws graphs on pieces of paper. The automobile manufacturer uses a smoothly rising graph to show the steadily increasing number of cars he sells, and smiles contentedly. The stock market speculator studies a jagged graph, which shows the ups and downs in the price of a railroad stock, and gnaws his nails. A graph shows many important features of the data, almost at a glance, much more clearly than the table of numbers from which it is drawn.

But the scientist is probably the most persistent user of the device. Already the graph of the solubilities of three different salts of sodium in Figure 21 has provided a typical illustration. You will find solubility diagrams a good form for practicing this aid to thought.

The Proportion of Salt to Water

Notice first an interesting way of looking at any such diagram as Figure 22. Here is a plane two-dimensional piece of paper with a single line, which represents the solubility of some salt in water, drawn across it. But any point on the piece of paper, whether that point is on the line or not, corresponds to some imaginable solution of the salt in a definite amount of water. The position of the point in the vertical direction on the paper specifies the amount of solid dissolved in the water, and the position in the horizontal direction specifies the temperature of the solution. If you change the temperature, the point

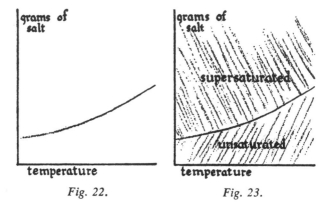

Fig. 22.   Fig. 23.

moves horizontally (to the right for a temperature rise). If you change the amount of solid dissolved, leaving the amount of water fixed, the point moves vertically.

Figure 23 emphasizes that the solubility *curve* separates the diagram into two *regions:* an unsaturated region and a supersaturated region. In general the curve runs across the paper diagonally—usually upward from left to right, because the solubility of most salts increases with temperature. Common salt is an exception in which the curve runs horizontally. Whenever the curve runs diagonally, you might think that there are two quite different ways of moving a point from one of the two regions to the other: it can be moved horizontally or vertically.

Now look at what you would actually have to do to a

solution in order to move the point representing it. Suppose you want to move the point A of Figure 24, which is in the unsaturated region, to the position B, in the supersaturated region. You reduce the temperature from T_A to T_B; and if no salt separates out, the job is done.

But suppose, instead, you want to move from A to the position E in Figure 25. If you try to move the point vertically, by adding salt to the solution at constant temperature, the salt will dissolve until you finally reach the

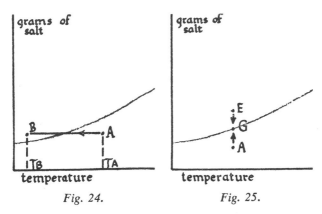

Fig. 24.   Fig. 25.

point G on the solubility curve, and there you will stick fast. Indeed, if you were trying to move in the other direction—to move the point E vertically down to the position A—you would encounter the same difficulty. You could let solid separate out of the solution until you got stuck at G. Thus you can think of the solubility line in these diagrams as a line which can be crossed horizontally but forms a barrier to vertical motion which at first sight seems impassable.

Any enterprising person, confronted with an impassable barrier, looks for a way around it, and here the way is clear. Take several steps to the side, so that you can move vertically in the unsaturated region, and then cross the barrier horizontally. Warm the solution until it is at the point C in Figure 26, dissolve salt in it until it is at D, and finally cool it again to attain the point E.

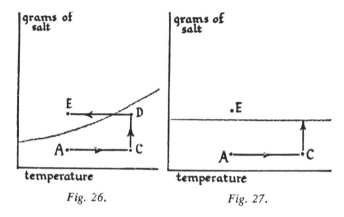

Fig. 26.   Fig. 27.

From this argument you might deduce that for common salt, or in any other case where the barrier is horizontal, the two regions are forever inaccessible from one another. Horizontal motion—change of temperature—can at best carry the point along the barrier in Figure 27, not across it. But thinking of how crystals of common salt are actually made by evaporating water from the solution, you see that the difficulty arises from the way these diagrams are drawn. They have been drawn for a fixed amount of solvent.

One way out of the difficulty would be to draw a diagram such as Figure 28, showing a family of curves instead of a single curve. Each member of the family belongs with a different fixed amount of solvent. No new measurement would be necessary to do this. The new curves could all be calculated from the old curve, using the principle that half as much solvent, for example, will dissolve half as much salt.

Then the family of curves shows that a solution that

started out unsaturated could be made supersaturated by evaporating part of the solvent, say half, and leaving the amount of solid fixed. As Figure 29 shows, the point representing the solution in the diagram does not move; it is the curve that moves while the point stays still. Before evaporation the point is on the unsaturated side of the appropriate barrier; after evaporation the point finds itself on the supersaturated side of the new barrier.

But notice that this whole family of curves is not really

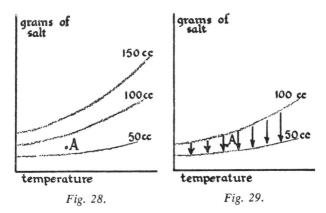

Fig. 28.   Fig. 29.

necessary after all. You see that they ought not to be necessary when you remember that no really new information went into them. All the information was contained in the one curve from which the others were calculated. It is only necessary to keep the one original curve, and make a slightly different interpretation of the vertical scale in the diagram. Instead of thinking of a fixed amount of water with varying amounts of salt dissolved in it, think of amounts of salt *per unit* amount of water, as in Figure 30, and choose for the unit amount the fixed amount that was used before.

Now the old diagram, with the new interpretation of its vertical scale, will serve also when water is allowed to

evaporate from the solution. Starting with unit amount
of water, you evaporate, say, half of it. Then the same
amount of salt is still in solution in half a unit amount of
water—that is to say, there is twice as much salt per unit
amount of water. The point that represents the solution
has risen vertically to a position twice as high as its
original position, as Figure 31 shows. In other words,
in this diagram you can proceed vertically by changing

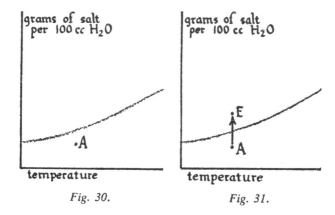

Fig. 30.   Fig. 31.

the amount of water as well as by changing the amount
of salt. You concern yourself with what really matters
—the proportion of salt to water. The vertical impassabil-
ity of the barrier has vanished.

A diagram like this offers a method for summarizing
on paper—by points, lines, and arrows—a succession of
operations whose description requires a great many
words. It makes many of the relationships between the
operations clearer than words could.

Salts That Form Hydrates

In drawing solubility diagrams for salts that form several different hydrates, some interesting new problems

arise, and with them come new opportunities for making physical processes clear diagrammatically. Nickel sulfate is a good example to examine. As you saw in Chapter II, it forms two hydrates, one with six molecules of water per nickel ion (the hexahydrate) and the other with seven molecules of water (the heptahydrate).

The first question to settle is: What should the vertical scale on the diagram represent? The best answer is to let it represent the amount of the salt the two hydrates and the solution all have in common—nickel sulfate. The solution contains nickel sulfate and water, and a line drawn in the diagram will represent the limiting proportion of nickel sulfate which the water will hold in solution at various temperatures.

Now recall the discussion in Chapter II of how to decide when the solution contains that limiting proportion. If a seed of the salt neither grows nor dissolves in the solution, the solution is "saturated" and contains the limiting proportion. But for nickel sulfate solutions there are two kinds of seeds which could be used, the hexahydrate and the heptahydrate.

You might, and in general you will, get two different answers for the limiting amount according to which kind of seed you use. The best way to proceed is to make two series of measurements, one using hexahydrate seeds and the other using heptahydrate seeds. Plotting the measurements, you will get two curves, which you can label with the names of the hydrates to which they belong, as in Figure 32.

But you will have some experimental difficulty making good measurements to use in plotting the high-temperature part of the heptahydrate curve and the low-temperature part of the hexahydrate curve. The diagram shows why these difficulties arise: the two curves cross each other at the temperature 30.7° centigrade. Examine the representative point A in Figure 33, which falls on the solubility curve for the hexahydrate at a temperature below 30.7° centigrade. It is in the supersaturated region of the diagram marked off by the curve for the heptahydrate. Hence, if any heptahydrate seeds are present they will grow, even though the hexahydrate seed does

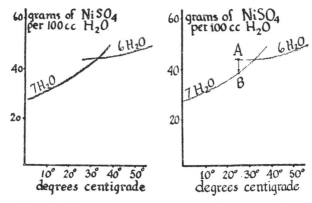

Fig. 32. TWO DIFFERENT CURVES *will represent the solubility of nickel sulfate, according to whether you test for saturation with a seed of the heptahydrate* (7H₂O) *or a seed of the hexahydrate* (6H₂O) *of the salt.*

Fig. 33. SEEDING WITH THE HEPTAHYDRATE *of nickel sulfate starts the representative point A of the solution moving toward B; a seed of the hexahydrate does not.*

not grow, until the representative point has moved down to B on the solubility curve for the heptahydrate. At the same time, as soon as the point moves down from A, it falls in the unsaturated region of the diagram marked off by the curve for the hexahydrate, and the seed of hexahydrate will begin to dissolve.

The most natural meaning of "the solubility curve of nickel sulfate," therefore, is the solid curve in Figure 34. In the low-temperature region it follows the heptahydrate curve, in the high-temperature region it follows the hexahydrate curve, and at 30.7° centigrade it changes slope abruptly. Careful experimental work may enable you to make measurements which provide the dotted portions of the curves. It is much easier to extend the hexahydrate curve than the heptahydrate curve by actual

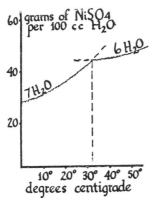

Fig. 34. THE SOLUBILITY OF NICKEL SULFATE *in any region of temperature is the lowest curve in the region. It changes slope, therefore, where two curves cross.*

measurements. The procedure we have given for growing nickel sulfate hexahydrate at room temperature depends on the fact that supersaturated solutions of nickel sulfate often do not deposit seeds of the heptahydrate spontaneously.

The temperatures given in Figure 19 for three different salts which form several hydrates are analogous to the temperature 30.7° centigrade for nickel sulfate. They are temperatures at which the solubility curves for the two hydrates cross each other.

PROBLEM 1

Plot the tabulated data for the solubility of sodium sulfate, in a diagram analogous to Figure 34, using such information in Figures 19 and 21 as you find helpful.

Temperature (degrees centigrade)	Solubility (g.Na$_2$SO$_4$ per 100 g.H$_2$O)	Temperature (degrees centigrade)	Solubility (g.Na$_2$SO$_4$ per 100 g.H$_2$O)
0.70	4.71	33.50	49.39
10.25	9.21	38.15	48.47
15.65	14.07	44.85	47.49
24.90	27.67	60.10	45.22
27.65	34.05	75.05	43.59
30.20	41.78	89.85	42.67
31.95	47.98	101.90	42.18

CHAPTER IV

Two Methods for Growing Crystals

Two general procedures for growing large single crystals of salts can conveniently be used at home. In both methods you suspend a seed crystal by a thread in a Mason jar containing the solution. In one, the "sealed-jar method," you supersaturate the solution and seal the jar to keep water from evaporating. The seed will grow as excess salt in the solution slowly crystallizes on it. This is the quickest and most useful way of growing most of the substances mentioned later.

In the other method, the "evaporation method," you start with a saturated solution and permit it to evaporate slowly. You leave the jar unsealed, and cover the top with a piece of cloth, both to reduce the rate of evaporation and to keep dust out of the solution. As water evaporates, the solution becomes supersaturated and the seed grows.

In both methods fairly constant temperatures are quite important, because changes in temperature change the degree of supersaturation. Consequently, it is wise to keep the jar somewhere in the house, possibly the basement, where it will not be disturbed and where the temperature varies the least.

Preparing a Saturated Solution

In both methods of growing crystals the first step is to make a solution that is saturated at the temperature at which the crystals will be growing. In the evaporation method you will then let the solution evaporate slowly after you have hung a seed in it. In the sealed-jar method you will heat the saturated solution to a higher temperature, where it is unsaturated. Then you will dissolve a little more salt in it, hang a seed in it, and cool it to the original temperature, where it will find itself supersaturated.

To prepare the saturated solution, you could proceed either by dissolving solid in an unsaturated solution or by withdrawing solid from a supersaturated solution. Notice now why the latter is the better procedure.

A solid salt at the bottom of a jar of water will dissolve quickly at first, but it will soon be surrounded by a concentrated solution. Since the solution is denser than the water, it will tend to stay at the bottom. If you do not stir the solution, further progress toward saturation will depend on diffusion of the salt upward into the more unsaturated part of the solution, a very slow process. If you stir the solution, you soon meet another problem. As the solution comes closer to the saturation point, the solid dissolves more slowly. The procedure needs a lot of attention over a long time.

A better procedure is to reach saturation by letting a supersaturated solution deposit its excess solid, as shown in Figure 35. The crystals at the bottom of the jar take solid out of the liquid, leaving the solution around them less dense than the rest of it. The less dense solution rises and a more concentrated solution replaces it in contact with the crystal surfaces. Thus the solution stirs itself, so to speak. Shaking the solution occasionally will speed

the process; but in time the solution will reach saturation
even without agitation.

The recipes in the next chapter give quantities of salt
and of water that produce solutions supersaturated at
temperatures below about 27° centigrade. These quantities have been worked out, whenever possible, to use a
full jar of the salt as suppliers usually package it. Make

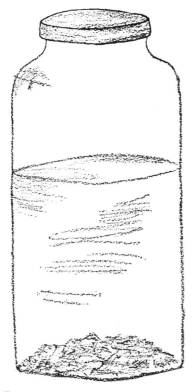

Fig. 35. TO MAKE A SATURATED SOLUTION, *seed a supersaturated solution and shake twice daily. It will put
its excess on the seeds and become saturated in two or
three days.*

the water measurements as accurately as you can, prefer-ably in cubic centimeters. The appendix shows how to convert temperatures from the centigrade to the Fahrenheit scale.

By heating the mixture of the salt and water to about 50° centigrade, you will be able to dissolve the salt in the water quite rapidly if you stir the mixture occasionally. Do not use aluminum vessels; they may be attacked by some of the solutions. A suitable vessel is a stainless steel double boiler, or a Mason jar placed in a saucepan of hot water. Keep a lid on the vessel between stirrings to reduce loss of water by evaporation from the solution.

When the salt has dissolved, pour the hot solution into a one-quart Mason jar and seal it to prevent evaporation. Then cool the solution to the temperature at which you expect to grow crystals. Since the solution is now supersaturated, seed it with a pinch of the salt to provide a place for the excess salt in the solution to deposit. Suitable seeds for this purpose are the crystalline powder left in the supply jar, or the powder left after evaporating a drop of the solution to dryness.

Seal the jar again, shake it well, and keep the solution at your expected growing temperature for at least two days; shake it twice a day and give it time to become saturated. If the temperature varies much, place the sealed jar in a pail of water. Since the water rises and falls in temperature more slowly than the surrounding air, it will act as a "thermal ballast," reducing the temperature fluctuations of the immersed jar.

When the precipitate stops growing, the solution has reached its saturation point. Pour off the clear solution into another container, taking care that the solution carries with it the least possible amount of the salt at the bottom of the jar. Then scrape that deposited salt onto a saucer, and when it has dried, return it to the supply bottle. Wash and drain the Mason jar, pour the saturated solution back into it, and seal it, since further evaporation would make the solution supersaturated in a short time.

Preparing a Seed Crystal

Any fragment of the solid, no matter how tiny, is a potential seed. But in order to be conveniently suspended by a thread, a seed must be ⅛ to ¼ inch long. Furthermore, it must be a single crystal so that the crystal growing from it will also be single.

You can prepare such seeds by pouring an ounce of your saturated solution into a small glass and setting it in an undisturbed place. As the solution evaporates, a few crystals will usually begin to grow on the bottom of the glass (Figure 36). If it becomes supersaturated without depositing crystals, add a very small amount of

Fig. 36. To grow seed crystals, *allow an ounce of saturated solution in a glass to evaporate slowly, and remove and dry the crystals when they have reached the proper size.*

crystalline powder from the supply bottle, or of the powder left after evaporating a drop of solution. Look at the glass and its contents once or twice a day; harvest the seeds when they have grown large enough for convenient handling, but before they grow so large that they touch and interfere with one another. Pick out the good seeds with tweezers, or pour off the solution and dump all the seeds on a paper tissue, where you can dry them well.

Save all good seeds. Your first crystal-growing efforts may fail, and when they succeed you will probably want to grow several crystals of the same substance. Furthermore, good seed crystals are excellent subjects for determining the angles between crystal faces with the "reflecting goniometer" you will read about further on.

You will notice that one face of each of these little crystals is slightly concave. It is the face that rested on the bottom of the glass. Not much solution is able to get under such a face to make it grow. But slight vibrations of the glass let a tiny amount of liquid under the edges of the crystal, and those edges slowly grow until finally the hollow left at the center of the face becomes deep enough to be noticeable.

Preparing the Growing Solution for the Sealed-Jar
(Supercooling) Method

The sealed-jar method requires the preliminary preparation of a supersaturated solution from the saturated solution described at the beginning of this chapter. You will prepare that solution by dissolving more salt in the saturated solution at a higher temperature, and then cooling the solution. The proper degree of supersaturation varies with the behavior of each salt: how fast it can order itself into a crystal without faults, and how highly its solution can be supersaturated without depositing seeds spontaneously. Of course, the crystal you grow at constant temperature cannot become larger than the amount of salt you add to a solution originally saturated at that temperature.

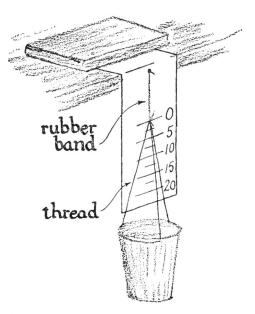

Fig. 37. BALANCES FOR WEIGHING SALTS. *In a rough spring balance, a thin rubber band can be used for the spring. Make a mark on the cardboard at the point where the rubber band and string meet when the paper cup is empty. Then mark for 2½ grams by putting a dime in the cup, and continue to about 20 grams. The rubber band will eventually show "fatigue"; the balance will not return to the zero mark when the cup is empty. Then you must replace the rubber band and re-calibrate the scale. A more accurate and permanent balance is fairly easy to make of scraps of wood, with razor blade suspensions. Study a manufactured balance for a pattern.*

The recipes in the next chapter, specifying amounts of salt to add to a saturated solution in order to super-saturate it, take these considerations into account. They are suitable for growing crystals in a room whose temperature fluctuates between 23° and 25° centigrade. If your growing conditions differ from this, you may find after trial that you get better results by slightly increasing or reducing the amount of added salt.

Weigh the amount of salt needed, put it in the double boiler, and pour the saturated solution over it. Figure 37 shows how to make a balance accurate enough for this purpose. Heat the solution slowly, stirring it until the salt has dissolved. In this operation, any tiny crystals that remained in the solution when you decanted it after saturating it will also be dissolved. Wash the Mason jar, and let it drain dry enough so that the amount of wash-water you will be adding to the solution is negligible. Pour the solution into the jar, seal the jar, and let the solution cool slowly.

Seeding the Solution in the Sealed-Jar Method

While the solution is cooling, prepare a cardboard disc to hold the thread for suspending the seed, as shown in Figure 38. Lay the seed on a piece of paper in preparation to tying it. Then wash your hands, for by this time there will be many invisible seed crystals on them. With clean hands, make a slip knot in a piece of sewing thread and tighten the loop around the seed, as shown in Figure 39. Attach the thread to the cardboard disc, and leave such a length of thread between seed and disc that the seed will be suspended an inch or two above the bottom of the jar.

Bring the solution to a temperature about three de-grees centigrade above the expected growing tempera-ture. If the solution is still warmer than that, put the

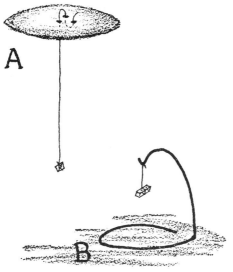

Fig. 38. SUSPENSIONS FOR SEEDS. *A—To make the disc for suspending the seed in a sealed jar, turn the jar upside down on a piece of cardboard, draw a circle around the mouth, and cut out the circular piece. This disc will fit over the top of the jar without falling in, yet permit the lid to be screwed down tightly. Make three small holes near the center of the disc, large enough for thread to go through. After the seed is tied, send the loose end of the thread through these holes, first up from the bottom, then down, then up again. The thread will shorten if you pull on the loose end. Adjust it so that the seed hangs about an inch above the bottom of the jar. B—A wire bent into a "cobra" is used to suspend the seed while crystals grow by evaporation. The important things to remember are that the top of the wire must be below the surface of the solution and that the base must be wide enough to prevent tipping.*

jar in a pail of cold water and stir the solution with a thermometer; if the solution is too cool, use hot water in the pail. In either circumstance stir the solution well to equalize the temperature throughout it.

Now you are ready to seed the solution. To plant one seed, excluding all others and preserving that one, takes care; it is one of the most critical steps in growing the

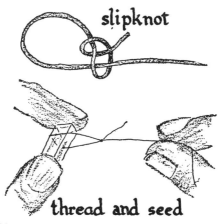

Fig. 39. To TIE THE SEED, *use common sewing thread with a slip knot.*

crystal. The seed you have prepared has other microscopic seeds on it; the air carries a dust of seeds. At the planting temperature, the solution is unsaturated. Soon after you have planted the prepared seed, all the microscopic seeds, and part of the prepared seed, will dissolve. Since the prepared seed is the largest, some of it will be left when the other seeds have dissolved completely.

As the seeded solution cools to the growing temperature, it will become supersaturated, and the prepared seed will stop dissolving and begin to grow. By watching the density currents around the crystal, shown in Figure 40, you can tell whether it is growing or dissolving.

Changes in Procedure for Growing by Evaporation

In the method of growing by evaporation, you change the foregoing procedure in some details. Heat the saturated solution to dissolve all unwanted seeds, return it to a clean jar, and cool it to a temperature a degree or two above the expected growing temperature, stirring it with a thermometer to make sure that the temperature is uniform throughout.

Tie the seed thread to a wire "cobra" (Figure 38B)

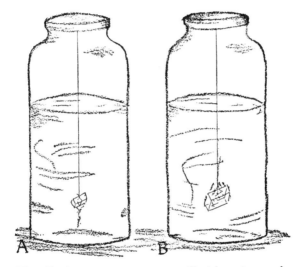

Fig. 40. A DENSITY CURRENT *will either rise or descend from a crystal hung in a solution if the solution is quiet. When the current descends (A) it is carrying extra salt dissolved from the crystal. Because the crystal is dissolving, the solution must be unsaturated. When the current ascends (B) it consists of a solution depleted of some of its salt. The lost salt has been deposited on the crystal; the crystal is growing, and the solution in the jar must be supersaturated.*

instead of a cardboard disc. Be sure that the cobra and the thread do not extend above the surface of the solution. A thread projecting from the solution will act as a wick. In this method, unlike the sealed-jar method, water will evaporate from such a wick, additional seeds will form on it, and there is danger that they will drop on the growing crystal. In any event, some unwanted seeds will drop from the surface of the solution as the water evaporates, but they cause trouble less often.

Place the cobra, with its suspended seed crystal, in the slightly unsaturated solution, and cover the top of the jar with a cloth held in place by a rubber band, as shown in Plate 12. Again growth will start when the solution cools to the growing temperature. The rate of growth thereafter depends on the rate at which water evaporates from the solution.

Growing and Harvesting a Crystal

Do not disturb the crystal during growth. Try to keep the temperature fairly constant, using a bucket of water for thermal ballast if necessary. By the sealed-jar method the crystal should grow to a good size in from three to six days. Some crystal "debris" may accumulate in the bottom of the jar, growing from unwanted seeds, but the debris will cause no damage so long as the desired crystal continues to grow without interference, as the crystal shown in Plate 13 is growing.

The crystals pictured later with the recipes have all been grown by these methods; the photographs will help you to decide when to harvest your crystals. When a crystal has attained full growth, pull it out and dry it immediately with a paper tissue or a soft cloth. Be careful how you handle it, especially if you intend to use it in the optical experiments described later, for it is soluble in water, and perspiration on your hands will damage its clear, plane faces. The best way to store it is to wrap it in a scrap of cloth and put it in a screw-topped jar to keep it from damage in either too dry or too humid air.

To grow a second crystal of the same substance, weigh the total amount of solid material you took out of the jar—both the crystal and the debris at the bottom. Then dissolve that weight of material in the old solution, warming and stirring to make the new growing solution.

An Example of a Recipe

At this point an example will help to make the foregoing procedures clear. The recipe is for growing a crystal of Rochelle Salt. You will find how to convert pounds into grams, and fluid ounces into cubic centimeters, in the Appendix.

  I. Supersaturated solution:
        a. 1 pound Rochelle Salt in 349 cc. (11.5 oz.) water
        b. 130 grams Rochelle Salt per 100 cc. water.
 II. Add to saturated solution:
        a. 31 grams Rochelle Salt
        b. 9 grams per original 100 cc. water.

The "a" quantities are based on your buying a certain quantity of the salt. They always specify the smallest amount of salt that will give satisfactory results, and they will spare you much weighing. The "b" quantities are ratios; from them you can calculate any desired quantity of solution.

To grow a Rochelle Salt crystal, dissolve one pound of the salt in 349 cc. of water, measured by a graduated cylinder, or 11½ ounces of water, measured by a kitchen measuring cup. Heat the mixture to dissolve the salt, seal it in a Mason jar, and let it cool. Then add some grains of Rochelle Salt; the supersaturated solution will deposit its excess on the added grains, and in a couple of days the solution will become saturated. Pour off the solution, and grow some seeds from an ounce of it.

Now you are ready to make the growing solution. To grow by evaporation, warm the saturated solution to dissolve the unwanted seeds, then let it cool again. To grow by the sealed-jar method, warm the saturated solution, add the "a" quantity of salt—31 grams—given in Part II of the recipe, and dissolve it. This is the growing solution in which you plant the seed, using the cardboard disc to suspend it.

Comparison of the Two Methods

Each method—sealed-jar and evaporation—has advantages and disadvantages. Both, of course, provide the indispensable condition for a crystal to grow from solution, supersaturation. In the sealed-jar method, the supersaturation arises from supercooling the solution. In the other method, evaporation of some of the water provides a progressive supersaturation of the solution.

Growing a crystal by evaporation, you can, at least in principle, get back almost all the dissolved solid in the form of a single crystal. But the rate of evaporation is hard to control: it depends on how humid the environment is, and how effectively casual drafts remove the evaporated moisture. Since evaporation occurs at the surface of the solution, the degree of supersaturation tends to be greatest there; unwanted seeds often form at the surface and may drop on the desired crystal. Any droplets of solution splashed on the sides of the jar, at the time it was filled, will evaporate to dryness, and the residue of crystalline dust may drop into the solution, providing a host of nuclei for crystallization.

On the other hand, supersaturating the solution by cooling it below the saturation temperature provides a control of supersaturation as good as the control of temperature. Often you can cool the solution three or more degrees centigrade below the saturation temperature without causing additional seeds to form spontaneously. Then a crystal can be grown for as long as a week at constant temperature. As the crystal grows, the supersaturation declines, and thus automatically provides the slower growth rate usually desirable as a crystal becomes larger. But the amount of material that can be deposited from the solution is clearly limited, even if you reduce the temperature again after the initial supersaturation has been exhausted.

PROBLEM 2

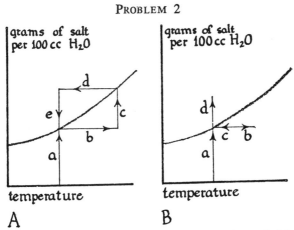

Identify which of these diagrams describes the sealed-jar method, and which the evaporation method, of growing crystals. Identify each of the lettered steps in each process.

PROBLEM 3

Explain what changes you would make in the sealed-jar method for growing crystals in order to adapt it for growing a crystal of anhydrous sodium sulfate.

CHAPTER V

Twelve Recipes for Growing Crystals

Using the following recipes, and the general procedures described in the preceding chapter, the authors of this book grew the crystals pictured in this chapter. The first seven recipes are described in greater detail than the last five. The substances to be crystallized in the first nine recipes can be purchased from sources given in the Appendix. The last three recipes require you first to prepare the substances by carrying out simple chemical operations on other materials which you can obtain from the same sources.

These recipes will enable you to start a many-colored crystal collection, to which you can add natural minerals and other artificially grown crystals. But the particular substances which the recipes cover have been chosen primarily because you can also use the products to study some interesting geometrical, mechanical, electrical, and optical properties of crystals.

At the head of each recipe, you will find the chemical name of the substance; a familiar name, if it has one; its chemical formula; its color; and the "crystal system" to which the crystals belong. The six crystal systems form the most important part of the crystallographer's scheme for classifying crystals; we shall get to them later in the book.

Potassium Aluminum Sulfate Dodecahydrate (Plate 14)

("Alum" or "Potash Alum"), $KAl(SO_4)_2 \cdot 12H_2O$, colorless, cubic system

   I. Supersaturated solution:
       a. 4 ounces alum in 567 cc. (19 oz.) water
       b. 20 grams per 100 cc. water.
  II. Add to saturated solution:
       a. 22 grams alum
       b. 4 grams per original 100 cc. water.

Alum grows well, offering no special problems, and is therefore the best choice for a first attempt at crystal growing. If the crystal grows too quickly, it will have many "veils" and look milky. If it does, in your second attempt reduce the II quantity in the recipe. The easiest way to make this change is first to harvest the veiled crystal and any debris in the jar, dry them, and weigh them. If the veiled crystal and debris together weigh *more* than the II quantity in the recipe, your "saturated" solution was supersaturated. A convenient way to re-make it is to break up the veiled crystal, giving it extra surfaces, and put it back in the solution. By letting these crystal fragments remain for several more days than you did the first time, you can make sure all the excess material deposits. If the crystal and debris weigh *less* than the II quantity, that quantity is too large to give satisfactory results at your growing temperature. Reduce the II quantity a little by redissolving only part of what you took out. Then you can seed the solution again.

Potassium Chromium Sulfate Dodecahydrate (Plate 15)

("Chrome Alum"), $KCr(SO_4)_2 \cdot 12H_2O$,

purple, cubic system

I. Supersaturated solution:
    a. 4 ounces chrome alum in 189 cc. (6⅔ oz.) water
    b. 60 grams per 100 cc. water.

II. Add to saturated solution:
    a. 9½ grams chrome alum
    b. 5 grams per original 100 cc. water.

Like ordinary alum, chrome alum grows well, but a saturated solution is so dark that you cannot see what is happening. Growing "mixed crystals" of ordinary alum and chrome alum is more satisfactory. Make a solution of chrome alum using 100 cc. of water. Pour this solution into a growing solution of ordinary alum slowly, while you hold the solution up to a light. You can mix the two in any quantities you wish, but for convenience stop adding the chrome alum while you can still see through the solution. Notice that the chrome alum solution is a dark blue-green, but the crystals are purple.

For the same reason that these mixed alum crystals will grow, you can grow one alum over another. For example, instead of harvesting a mixed alum crystal, you can lift it out—thread and all—and use it as a large "seed" on which to obtain a further growth of ordinary alum. The final result will be a clear crystal with a lavender crystal inside it.

Over a long time, you can grow a very large crystal of any of these alums by the evaporation method. Be sure that the top of the cobra is well under the water surface. Evaporation should be slow, and the crystal should be inspected every few days. If the batch of solution is large enough, you can grow the crystal for several months.

Potassium Sodium Tartrate Tetrahydrate (Plate 16)

("Rochelle Salt"), $KNaC_4H_4O_6 \cdot 4H_2O$, colorless,
orthorhombic system

I.   Supersaturated solution:
    a. 1 pound Rochelle Salt in 349 cc. (11.5 oz.) water
    b. 130 grams per 100 cc. water.
II.   Add to saturated solution:
    a. 31 grams Rochelle Salt
    b. 9 grams per original 100 cc. water.

Next to alum, Rochelle Salt is the substance best suited for an initial attempt at crystal growing. In growing Rochelle Salt, the seed solution often supersaturates without depositing any seeds. Should no seed appear within two days, add a very few Rochelle Salt particles from the supply bottle. Once started, the seeds will probably grow very quickly, and therefore you should inspect them twice a day.

The fact that the solubility of Rochelle Salt varies greatly with temperature leads to another problem you may meet. You must be especially careful to observe and control the temperatures at which you work; it is easy to lose the planted seed by dissolving it off the thread before the solution becomes supersaturated.

Because the seeds and the final crystal dehydrate easily, it is best to keep them in a closed jar, wrapped in a scrap of cloth or cotton batting. Moistening the packing with a drop of the Rochelle Salt solution will furnish added protection.

Oddly enough, two different modifications of shape, shown (Plate 16) to the right of the normal Rochelle Salt crystal, are obtainable by adding copper ions to the growing solution. Five cubic centimeters of a saturated solution of copper acetate (one gram of copper acetate in 10 cc. of water) added to a growing solution of Rochelle Salt made from 100 cc. of water (Ib and IIb above) will provide a solution which grows long, thin crystals. That last solution, with one pellet of sodium hydroxide dissolved in it, will grow flat, wide crystals.

Two Hydrates of Nickel Sulfate

Nickel Sulfate Hexahydrate (Plate 17A)
$NiSO_4 \cdot 6H_2O$, blue-green, tetragonal system
Nickel Sulfate Heptahydrate (Plate 17B)
$NiSO_4 \cdot 7H_2O$, green, orthorhombic system

I.   Supersaturated solution:
    a. 1 pound $NiSO_4 \cdot 6H_2O$ in 394 cc. (13.8 oz.) water
    b. 115 grams $NiSO_4 \cdot 6H_2O$ per 100 cc. water.
For $NiSO_4 \cdot 6H_2O$ (nickel sulfate hexahydrate)
II. Add to saturated solution:
    a. 76 grams $NiSO_4 \cdot 6H_2O$ (or 81 grams $NiSO_4 \cdot 7H_2O$)
    b. 19 grams $NiSO_4 \cdot 6H_2O$ per original 100 cc. water.
For $NiSO_4 \cdot 7H_2O$ (nickel sulfate heptahydrate)
III. Add to saturated solution:
    a. 27½ grams $NiSO_4 \cdot 6H_2O$
    b. 7 grams $NiSO_4 \cdot 6H_2O$ per original 100 cc. water.
263 grams $NiSO_4 \cdot 6H_2O$ has the salt content of 281 grams $NiSO_4 \cdot 7H_2O$.

Although both these hydrates of nickel sulfate grow well, you must take a few extra precautions in order to grow the particular hydrate you want. In preparing seeds at room temperature the heptahydrate ($NiSO_4 \cdot 7H_2O$)

will usually come out of solution. The first problem is therefore to produce a suitable seed of the hexahydrate ($NiSO_4 \cdot 6H_2O$). This is best done by seeding an ounce of the supersaturated solution (Solution I), before it is made into a saturated solution, with a grain or two from the supply bottle, which contains the hexahydrate. If the seeds growing from these grains are chunky instead of long, they are crystals of the hexahydrate (Plate 17).

In making the saturated solution from Solution I, do not use seeds from the salt in the supply bottle. Instead, let the solution cool to room temperature and then shake it to start the excess coming out. If no seed forms, let a drop evaporate to dryness and seed with the residue. This is necessary because the saturation point of the solution differs for the two hydrates, and the recipe is adjusted for saturation with respect to the heptahydrate, not the hexahydrate.

When you decant the saturated solution from the excess salt, observe the appearance of the salt, making sure that it is the heptahydrate and not the hexahydrate. Then put it in a properly labeled bottle. Either hydrate can be added to the solution, whenever the recipe calls for dissolving additional salt, but the weight to be added depends on which hydrate you use.

The seeds for making heptahydrate crystals grow well by the usual method, evaporation of an ounce of saturated solution. Since the heptahydrate dehydrates easily, protect the crystals by keeping them in a sealed jar containing a little powdered heptahydrate. Despite their tendency to dehydrate, crystals of the heptahydrate occasionally occur in nature; the mineralogist calls them "morenosite."

There is a great difference in the shapes of the crystals of these two hydrates. There is even some difference in their colors: the hexahydrate is a bluer green than the heptahydrate. Although you cannot grow a single crystal of the anhydrous sulfate, you can easily make some in powdered form to see its color. Put a little of the powder from either of the supply bottles on the blade of an old knife and heat it in a candle flame or over the stove. Water will boil out, and after all the bubbling stops, a yellow powder will remain. In order to satisfy yourself that this powder is still nickel sulfate, dissolve it in some water and then let the water evaporate; you will recognize the hydrated salt. The dehydrated salt will dissolve quite slowly, because it must become hydrated again before it dissolves.

PROBLEM 4

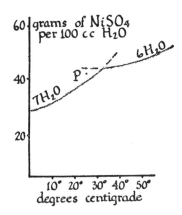

Describe the simplest experiment you can think of which would show unquestionably that it is possible to make a solution of nickel sulfate in water (such as that represented by the point P above) which is supersaturated with respect to the heptahydrate and unsaturated with respect to the hexahydrate.

Sodium Bromate (Plate 18)

$NaBrO_3$, colorless, cubic system
I.   Supersaturated solution:
     a. 8 ounces sodium bromate in 454 cc. (15.4 oz.) water
     b. 50 grams per 100 cc. water.
II. Add to saturated solution:
     a. 9 grams sodium bromate
     b. 2 grams per original 100 cc. water.

Sodium bromate crystals tend to develop veils during growth, unless they are grown quite slowly. Moreover, some debris usually forms on the bottom of the jar. The crystals have a tetrahedral shape.

Sodium Chlorate (Plate 19)

$NaClO_3$, colorless, cubic system
I.   Supersaturated solution:
     a. 1 pound sodium chlorate in 400 cc. (13.5 oz.) water
     b. 113.4 grams (¼ lb.) per 100 cc. water.
II. Add to saturated solution:
     a. 16 grams sodium chlorate
     b. 4 grams per original 100 cc. water.

Sodium chlorate crystals grow easily in the form of cubes. A crystal of sodium chlorate has the same arrangement of atoms as a crystal of sodium bromate, but with chlorate ions replacing the bromate ions. The fact that sodium bromate grows in tetrahedra and sodium chlorate grows in cubes shows that small differences in atomic character can make large differences in crystal growth.

You can change the growth shape of sodium chlorate by adding some borax to the solution. The borax does not become part of the crystal; it merely makes the tetrahedral faces appear on the crystal. In order to suppress the cube faces completely, you must add 6 grams of borax for every 100 grams of sodium chlorate in the growing solution; in the IIa solution this requires 25 grams of borax.

You can seed a sodium chlorate solution containing borax with a cubical seed of sodium chlorate, and see the seed grow into a tetrahedron. Conversely, tetrahedral seeds of sodium chlorate grown from a solution containing borax will grow into cubes in a pure solution of sodium chlorate.

Sodium Nitrate (Plate 20)

("Niter" and "Saltpeter" are confusingly used for both sodium nitrate and potassium nitrate.)

$NaNO_3$, colorless, hexagonal system

I. Supersaturated solution:
   a. 1 pound sodium nitrate in 412 cc. (14 oz.) water
   b. 110 grams per 100 cc. water.
II. Add to saturated solution:
   a. 12 grams sodium nitrate
   b. 3 grams per original 100 cc. water.

Sodium nitrate is not as easy to grow as the substances previously mentioned; but it forms a very interesting crystal, sharing many properties with the mineral calcite and useful in the studies of cleavage, glide, and double refraction you will read about further along. The range of supersaturation within which it grows well is very narrow, and the growth is therefore quite sensitive to temperature changes.

Seed it first at your usual seeding temperature. Should the seed dissolve, reseed the solution at a slightly lower temperature. If the solution deposits too many unwanted seeds during the growth of the crystal, lower the IIa quantity of the recipe by weighing the crystal and debris, and redissolving only part of the total. At best, the top part of the crystal will contain veils, but you will be able to cleave these off the harvested crystal.

Potassium Ferricyanide (Plate 21)

("Red Prussiate of Potash"), $K_3Fe(CN)_6$, red, monoclinic system

Dissolve 93 grams of potassium ferricyanide in 200 cc. of warm water, cover the solution, and allow it to cool. Do not be especially afraid of the word "cyanide" in the name; this substance is no more poisonous (and no less!) than the others in this list.

Copper Acetate Monohydrate (Plate 22)

$Cu(CH_3COO)_2·H_2O$, blue-green, monoclinic system

Dissolve 20 grams of copper acetate monohydrate in 200 cc. of hot water. If a scum of undissolved material persists, add a few drops of acetic acid and stir well. Cover this solution, and allow it to cool and stand for a few days; usually it will deposit crystals spontaneously, and they will grow large enough to use for an optical experiment described later. You can grow larger crystals by using these as seeds in the general procedures already described, but you will find it hard to grow them as large as you can grow the other crystals mentioned in this chapter.

Strontium Formate Dihydrate (Plate 23)

$Sr(HCOO)_2·2H_2O$, colorless, orthorhombic system

To 150 cc. of water, add 18.5 cc. of 88 per cent formic acid. Warm this solution, and add to it 29.5 grams of strontium carbonate, in small portions so that the foaming is not excessive. When all the carbonate has dissolved and bubbles of carbon dioxide no longer form in the solution, allow it to cool and deposit crystals.

Lithium Trisodium Chromate Hexahydrate (Plate 24)

$LiNa_3(CrO_4)_2 \cdot 6H_2O$, yellow, hexagonal system

Dissolve 60 grams of sodium dichromate dihydrate in 130 cc. of hot water, and add 7.5 grams of lithium carbonate in small portions. When the bubbling of carbon dioxide has stopped, add 8 grams of sodium hydroxide, and stir until it has dissolved. Cover the solution and let it cool and stand. If it fails to deposit crystals spontaneously within a day, add seeds obtained by evaporating a drop of it. Notice that, like alum, this substance is a "double compound."

Calcium Copper Acetate Hexahydrate (Plate 25)

$CaCu(CH_3COO)_4 \cdot 6H_2O$, blue, tetragonal system

Add 22.5 grams of powdered calcium oxide to 200 cc. of water, pour into the mixture 48 grams of glacial acetic acid, and stir until the solution is clear. If there is a small insoluble residue, filter the solution. Dissolve separately 20 grams of copper acetate monohydrate in 150 cc. of hot water. Mix the two solutions, cover the mixture, and allow it to cool and stand for a day. If it does not deposit crystals spontaneously, let a drop of the solution evaporate and scrape the resulting seeds into the bulk of the solution.

Notice that this solution contains four atoms of calcium for every atom of copper; whereas the crystals contain equal numbers of calcium and copper atoms. A solution containing the latter proportion deposits crystals of copper acetate monohydrate, not crystals of the double compound. Hence, you cannot recrystallize the double compound from pure water. But, of course, you can replenish a solution from which you have grown crystals of the double compound by redissolving an equal weight of that compound.

CHAPTER VI

Building Blocks for Crystals

Noticing that crystals grow, people many centuries ago jumped to the conclusion that they grow as animals do —from food taken in by them and assimilated to their substance. Occasionally someone noticing certain crystal behavior must have guessed that crystals grow from outside. But before the systematic practice of recording and reporting such observations came into use, mankind learned much and forgot it, relearned it and forgot again.

In time, accumulated reports of observations made clear the similarities and distinctions between crystals and living organisms. About three centuries ago people began to notice that they could think of any crystal as built up of tiny building blocks all alike—blocks like the sugar cubes from which the model of the alum crystal in Plate 5 is made, but too small to be individually distinguished. Crystal growth could then be pictured as the deposition of little identical building blocks, side by side and on top of one another. The evidence leading to this picture was drawn partly from observing the cleavage in calcite, which we shall come to later, and partly from studies of the shapes of crystals.

The Shapes of Crystals

In learning what you can about crystals from their external forms, the first step is to sort out in your mind what the forms of all crystals have in common. Notice, for example, that each of the crystals which you grow by the recipes in Chapter V is bounded by flat surfaces. These surfaces are called "faces"; wherever two faces meet, there is a straight "edge"; and three or more edges meet in a "corner."

Often the edges bounding a face make up a fairly simple plane figure—a triangle, or a square, or the like. And often the faces bounding the whole crystal make up a correspondingly simple solid figure—a cube, a tetrahedron, or an octahedron (Figure 41). Sometimes the

Fig. 41. POLYHEDRON *is the name given to a solid bounded by flat faces. Many simple polyhedra have specific names. A—The cube has six square faces. B—The regular tetrahedron has four equilateral triangular faces. An equilateral triangle has three equal sides. C—The regular octahedron has eight equilateral triangular faces. D—The rhombic dodecahedron has twelve rhombic faces. A rhombus has four equal sides.*

crystalline shapes are more complicated, but they all are "polyhedra"—figures bounded by a finite number of plane faces.

The faces of the alum crystal, for example, comprise essentially eight triangles, six little squares and twelve long rectangles. Looking at any alum crystal in detail, you will find that some of the faces of the same kind are slightly different in size, and that corners have been chopped off some of the squares and triangles. These variations occur because the solution from which the crystal grew varied in concentration from one point to another around the growing crystal, and the faces therefore grew at slightly different rates. Usually you will have no trouble visualizing an ideal shape, from which these little variations have been removed, as is done in Figure 42.

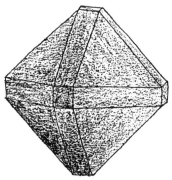

Fig. 42. AN "IDEAL" ALUM CRYSTAL *might be visualized as a combination of three polyhedra shown in Figure 41: a cube, an octahedron, and a rhombic dodecahedron, cutting off one another's corners and edges to form a polyhedron bounded by six little squares, eight triangles, and twelve long rectangles.*

The alum crystal is also a good example of how you can visualize a crystal as made up of a combination of simpler shapes. By extending all eight of the triangular faces of the alum crystal until they meet, you will eliminate the little squares and long rectangles and produce an octahedron. Since those triangular faces were the largest faces in the first place, you could roughly describe the crystal as having the shape of an octahedron.

If, instead of the triangular faces, you extend the six

little square faces until they meet, a cube appears. Extending the twelve long rectangles in the same way, you produce a rhombic dodecahedron, with twelve diamond-shaped faces, like that on the right side of Figure 41. You can think of the ideal shape of the alum crystal in Figure 42 as a combination of those three polyhedra—all three centered at the same place, and all three chopping off some of one another's corners and edges.

Crystallographers call the characteristic shape of a crystal its "habit." They might say, "The habit of alum is shown in Figure 42," or they might speak of alum as "having an octahedral habit."

The habit of a crystal—the relative sizes of its characteristic faces—is determined by the rates at which the different sorts of faces grow on it. Under ideal conditions, when all faces on a crystal are exposed to a solution with uniform concentration and temperature, all faces of the same sort—the six little square faces on an

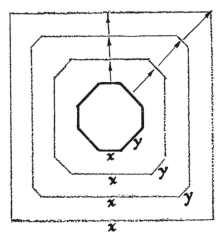

Fig. 43. SUCCESSIVE STAGES IN GROWTH *of an imaginary two-dimensional crystal. Growing faster than the x-faces, the y-faces may finally disappear.*

alum crystal, for example—will grow at the same rate. But, of course, faces of different sorts may still grow at different rates.

One consequence of the differences in the growth rates of different sorts of faces is the principle that *a crystal is surrounded by its slowest-growing faces.* This principle has a surprising sound, but a little thought about Figure 43 will convince you. The diagram shows successive stages in the growth of an imaginary two-dimensional crystal. The crystal starts with eight "faces," of two different sorts. The four y-faces are growing faster than the four x-faces, as you can see from the fact that the successive positions of any one y-face are farther apart than the successive positions of an x-face. The x-faces become larger and larger, while the y-faces become smaller and might eventually vanish by growing right out of the crystal.

Figure 44 exhibits the results of the same reasoning

Fig. 44. A GROWING ALUM CRYSTAL, *starting in the shape of a cube, might ultimately take the shape of an octahedron.*

applied to alum. If you went to the trouble to cut a cube-shaped crystal of alum, and planted it as a seed in a supersaturated alum solution, it would take on in succession the shapes diagrammed. The appearance of the crystal would become less and less cubical and more octahedral because the cube faces grow more rapidly than the octahedron faces.

This is an example of one kind of evidence of orderliness which you met in Chapter I. Here is a property of any crystal—its growth rate—which *varies with direction.* For the imaginary two-dimensional crystal of Figure 43, you can say that you know the growth rates in eight directions; the directions perpendicular to its eight faces. In fact, imagining the crystal as starting from a seed so small that it is invisible, you would represent the growth rates in the eight directions by arrows all starting at the same point, as in Figure 45. If the growth rates were the

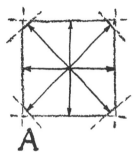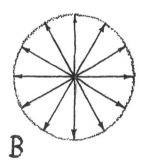

A   B

Fig. 45. CRYSTAL GROWTH RATES *in different directions can be diagrammed by arrows, starting from one point and having lengths proportional to the rates of growth. Then lines drawn perpendicular to the ends of the arrows (A) will enclose the characteristic shape which the crystal would take if it started from a tiny seed and grew constantly at those rates. If the growth rates were the same in all directions (B) the crystal would grow into a sphere.*

same in the eight directions, the eight arrows would all have the same length.

If the growth rates of this crystal were the same in *all* directions, you could represent that fact by drawing arrows of the same length in all directions from the center. Then, drawing faces perpendicular to all these arrows, you see that the shape of the crystal would be a circle. Correspondingly, if a real crystal grew equally fast in all directions from an invisible seed, it would grow into a sphere. Since no real crystal ever adopts a spherical shape when it is free to grow in its natural way, you can conclude that all crystals have at least one property—their growth rate—which is different in different directions.

PROBLEM 5

Draw or describe the shapes which would be taken by crystals whose relative growth rates in several directions were as described below. In all other directions the growth rates are so fast that the corresponding faces do not appear.

A. In a two-dimensional crystal, the rate in the four x-directions, at 90° to each other, and in the four y-directions, at 45° to the x-directions, are in the ratio $x/y = 5/4$.

B. In a three-dimensional crystal, the rates in one plane are as given in A, and the rates in the two directions perpendicular to that plane are twice as great as in the x-directions.

C. A three-dimensional crystal shows faces which can be described as resulting from a set of x-rates in the directions toward the eight corners of a cube, and a set of y-rates toward the mid-points of the twelve edges of the same cube. The rates have the ratio $x/y = 5/4$.

Distortions of Crystalline Shapes

A single crystal of quartz will never grow naturally as a sphere, but it is quite willing to be cut and polished into one, for the use of a crystal gazer. Indeed, there are many ways in which the shapes of crystals may be so distorted that their natural forms are almost unrecognizable.

When you grow crystals, you will surely encounter two of these ways. The most obvious one afflicts crystals growing on the bottom of your glass or Mason jar. While all other faces are growing, the face resting on the bottom of the container cannot. The resulting crystal will look somewhat like a properly grown crystal which has had a part removed.

A seed crystal, falling through a growing solution, will usually come to rest on one of its largest faces, because that is its most stable position under the pull of gravity. Consequently, the crystal grown from the seed will also be resting one of its largest natural faces. If it is alum, whose largest faces are the triangular faces of the octahedron, the crystal may look like a thick triangular tablet, as Figure 46 shows.

When cube faces also appear on the alum crystal, the tablet may acquire a hexagonal contour, such as you see

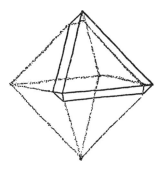

Fig. 46. An alum crystal *growing at the bottom of the jar cannot add material to the face on which it rests. It may look like the triangular tablet outlined here, instead of the octahedron.*

in Figure 47A. And if by chance the seed crystal has come to rest on one of its dodecahedron faces, the crystal may have the similar, though somewhat less symmetrical, contour shown in Figure 47B. The appearance of these crystals is far from suggesting the octahedral habit of alum, until you analyze how they arose. Even experienced crystallographers and microscopists can be confused by distortions having this simple origin.

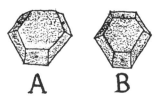

Fig. 47. TABLETS OF ALUM *growing at the bottom of the jar may have similar and confusing shapes if they rest (A) on an octahedron face and (B) on a rhombic dodecahedron face. Dots mark the cube faces in these diagrams.*

There is another distortion of the ideal shapes of crystals which you will surely observe. When a crystal is growing in a supersaturated solution, the crystal removes material from the solution, leaving it less dense and so setting up slowly moving "density currents" in it. Already you have found density currents useful, in preparing saturated solutions and in testing whether a solution is unsaturated or supersaturated. Notice now in Figure 48 how these currents can influence the shape of a growing crystal.

Figure 48A shows the ideal outline of an imaginary two-dimensional crystal. Placed in a supersaturated solution, it will begin to grow on all its faces, at rates represented by the arrows in Figure 48B. But the act of growth, withdrawing material from solution, sets up a density current, which climbs up the side of the crystal.

Plate 1. A cast slab of zinc, when broken apart, will often show the outlines of component crystals, because it has broken along boundaries between crystals.

Plate 2. Crystals of a metal can be grown in a test tube. Put a drop of mercury in a solution of 0.5 g. of silver nitrate in 20 cc. of water. Some of the mercury goes into solution, liberating metallic silver, which combines with the remaining mercury to form a crystalline alloy. The blade-shaped crystals will continue to grow for about a week.

Plate 3. Cleaving mica is easy. Push the point of a pin into the edge of a mica crystal; then work the pin back and forth along the edge, slowly pushing it at the same time.

Plate 4. Table salt often has natural faces showing on individual grains. Under a magnifying glass they appear as cubes.

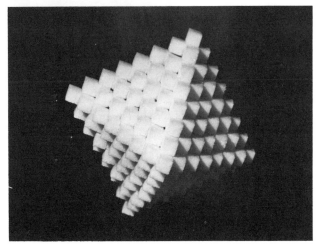

Plate 5. An alum crystal retains its characteristic symmetrical shape, no matter how large it grows. Because you can build this shape by piling up little cubes, you can think of an alum crystal as made of regularly arranged cubical blocks comparable to a molecule in size.

Plate 6. Cast brass doorknobs, etched by sweat over the years, often show the arrangement of component crystals.

Plate 7. To see crystals grow from a molten substance, put 25 g. of salol in a tightly closed bottle, and melt it by standing the bottle in hot water. When you let it cool, little crystals of salol will form in the molten material, and you can watch them get bigger until finally the salol has all solidified into a polycrystalline mass. To grow salol from solution, warm a mixture of 25 g. of salol and 100 cc. of denatured alcohol in a closed bottle, shaking it until the solution is clear. Allow the solution to cool, and if crystals do not form spontaneously, "seed" the solution with a bit of solid salol.

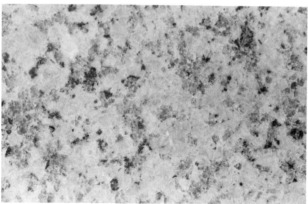

Plate 8. Granite, freshly broken, clearly shows its component crystals of clear or grayish quartz, pink feldspar, and black mica. Cut and polished, it is widely used as a facing stone on formal buildings. Its component crystals are easily seen under a hand lens.

Plate 9. Quartz, perhaps the commonest and most universally recognized mineral, occurs in large beautiful crystals in Brazil and Venezuela. Tiny rounded crystals of quartz form the principal ingredients of most sands (lower left), and when these are cemented together by other substances they form sandstone. Many of the rounded pebbles you find (lower right) are made of quartz.

Plate 10. A snow flake is a single crystal of water. The flat hexagonally symmetric flake is the least frequent type, but it is familiar because its beauty inspires pictures and imitations.

Plate 11. An ethylene diamine tartrate crystal with unexpected crystals growing on its surfaces. Later they were identified as crystals of the monohydrate.

Plate 12. Growing by evaporation.

Plate 13. Growing in a sealed jar.

Plate 14. Crystal of Potassium
Aluminum Sulfate Dodecahydrate
"Alum" or "Potash Alum" (colorless).

Plate 15. Crystal of Potassium Chromium Sulfate
Dodecahydrate "Chrome Alum" (purple).

Plate 16. Crystal of Potassium Sodium Tartrate Tetrahydrate
"Rochelle Salt" (colorless).

Plate 17A. Crystal of Nickel
Sulfate Hexahydrate (blue-green).

Plate 17B. Crystal of Nickel
Sulfate Heptahydrate (green).

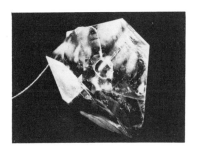

Plate 18. Crystal of Sodium
Bromate (colorless).

Plate 19. Crystal of Sodium Chlorate (colorless).

Plate 20. Crystal of Sodium Nitrate (colorless).

Plate 21. Crystal of Potassium
Ferricyanide (red).

Plate 22. Crystal of Copper Acetate
Monohydrate (blue-green).

As it climbs, it loses still more material to the crystal, and soon the solution bathing the crystal has a lower degree of supersaturation near the top of the crystal than at the bottom. The growth rates on faces near the top become less than the rates on corresponding faces at the bottom, and when a steady state of current flow has been established, the new growth rates could be rep-

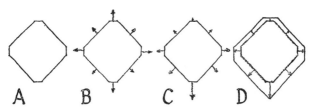

Fig. 48. UNEVEN GROWTH RATES *on similar faces make the faces look unlike. A crystal with two kinds of faces (A) would retain its shape if the supersaturation of the solution stayed uniform over all the faces and provided the growth rates shown at (B). But the supersaturation becomes lower near the top of the crystal, and the growth rates change to those shown at (C). The crystal then acquires a less symmetrical shape (D).*

resented by the arrows in Figure 48C. Finally the crystal will look like Figure 48D; corresponding faces will not have corresponding sizes.

You may ask, "If the growth rate varies with the degree of supersaturation, and the degree of supersaturation varies from top to bottom of a *single face,* how can all parts of that face continue to grow outward at the same rate?" In fact, it is extraordinary how well the various parts of a face seem to keep track of one another and succeed in arriving at compromises. But they cannot be pushed too far. As a face grows larger, it has increasing difficulty letting its left hand know what its right hand is doing, because they are farther apart.

Moreover, there is room for larger differences in the degree of supersaturation between its parts. If it is to remain unflawed, a crystal must be grown more slowly as it becomes larger.

The "hopper growth" shown in Plate 26 is a spectacular result of a variation of growing conditions over the parts of a crystal face. The schematic diagrams in Figure 49 will help you to see what has happened. The variation

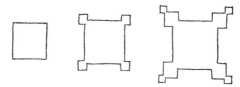

Fig. 49. IDEALIZATION OF HOPPER GROWTH. *Faster growth at corners and edges than on faces can be self-perpetuating when it partly shields the center of the face from a fresh supply of solution.*

in degree of supersaturation over a growing crystal often urges its edges to grow faster than the centers of its faces. An edge can draw upon more solution than a similar area on a face because the edge looks outward in more directions. If it succeeds in racing ahead of its surrounding faces, some little faces will grow out from it, and they will tend to shield the original face from fresh solution and so will further exaggerate the difference in supersaturation between the face and the edge.

Analysis of the shapes of snowflakes, such as that shown in Plate 10, often suggests that a similar succession of especially fast growths has occurred at the corners of a hexagonal plate. But the symmetry of these growths on a snowflake remains extraordinary; whatever one corner does, the others do too, at the same time and to the same extent. Several explanations have been offered for this unanimity in the detailed behavior of a snowflake in the six directions, but none has yet been generally accepted.

Besides these two ways—mechanical obstructions and variations of supersaturation—by which the natural growth shape of a crystal may be distorted, there is a third way, perhaps the most interesting of all. That way is a change in *chemical* environment. The recipes have already given you two instances in which you can observe this yourself: the habit of Rochelle Salt is changed when it grows in the presence of copper ions, and the presence of borax changes the habit of sodium chlorate from cubic to tetrahedral.

Clearly, in both changes, the effect of the substance added to the solution has been either to accelerate the growth of one sort of faces or to slow down the growth of another sort. For example, you could explain the effect of borax on sodium chlorate by saying either that borax increases the growth rate of the cube faces, or that borax reduces the growth rate of the tetrahedron faces. How can you decide which?

The changes of habit that copper ions produce in Rochelle Salt enable you to see with some clarity what is happening and to guess that similar action is occurring in many other crystal changes. If the modified Rochelle Salt crystals are clear, you will see that a thin layer of crystal at some of the faces has taken on a pale blue color, and that the other faces have the colorless transparency of pure Rochelle Salt. The colored faces are precisely the ones that would be affected if slower growth rates caused the change of habit.

You can guess that for some reason copper ions have attached themselves to those faces tenaciously and have not attached themselves to the other faces. Their presence becomes a partial barrier to the arrival and ordering of more molecules of Rochelle Salt, and thus reduces the usual growth rate. It is easy to visualize a mechanical blocking of this kind; it would be quite hard to visualize how an impurity could speed up the process of ordering molecules at the surface of a growing crystal.

Nevertheless, this explanation leaves much unexplained. Why do the copper ions prefer one set of faces to another? Why do they change their preference when you add sodium hydroxide to the solution? Why do they behave in this way with Rochelle Salt and not with, say, sodium chlorate?

There are many other instances of "adsorption"—the tenacious attachment of one substance to a surface of another. When the attachment shows a preference for certain crystal faces, it is called "selective adsorption." These phenomena are not easily explained, and they are so specific that the explanation has to be tailored to fit each one. Indeed, some adsorptions defy explanation at present.

Selective adsorption probably causes many of the habit modifications of naturally occurring minerals—for example, two different habits of natural quartz shown in Plate 27. Mineralogists have catalogued many habits of calcite.

The Angles Between the Natural Faces

Despite these distortions, one feature of the crystals remains constant. So long as the crystals are made of the same substance, the angle between any chosen pair of faces stays unchanged, unless change of habit has caused the faces to disappear. For example, even though the shapes of Rochelle Salt crystals vary from long and thin to short and fat, the angles between corresponding pairs of faces will be the same on them all. They are characteristic of the substance, unvarying from sample to sample of it. This *constancy of the interfacial angles* was the early crystallographers' first quantitative discovery.

Sometimes a chemist uses this fact to identify a substance. He may measure the angles on a crystal of an unknown material and compare them with the angles on a known crystal which he thinks is the same as his unknown. Or he may compare his angles with records of angles which have been compiled for known materials.

To measure angles, the earliest crystallographers used

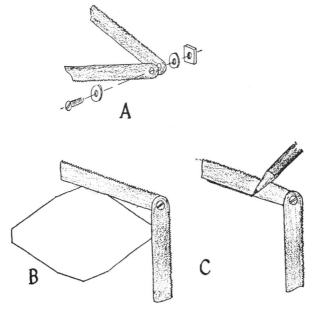

Fig. 50. A "CONTACT GONIOMETER" for measuring the angle between a pair of faces on a crystal is made (A) by breaking the blade of a hacksaw and screwing the two pieces together with a machine screw and nut, and two rubber washers. After adjusting it to a pair of faces on a crystal (B), the angle between the faces can be drawn on a piece of paper (C), and compared with the angle between similar pairs of faces on other crystals of the same material, to verify the "law of the constancy of interfacial angles."

a "contact goniometer." You can make a simple instrument like theirs from the blade of a hacksaw, as shown in Figure 50. After adjusting its jaws until they fit snugly against the two faces whose angle you want to measure, you can lay the goniometer on a piece of paper and use the jaws as rulers for a pencil. By comparing several crystals in this way, you can satisfy yourself that all the crystals you have grown of any one substance have the same interfacial angles.

You can use this instrument also to duplicate another discovery made by the early crystallographers: that any crystal can be regarded as made of a single type of building blocks, laid on top of one another and side by side in regular array. This was a discovery of prime importance, for it gave early evidence that matter is not a structureless jelly, but has an *ultimate fine structure*. Today crystallographers can identify an atom, a molecule, or a group of rather few atoms or molecules, with a building block, in a fashion we shall describe later.

Rochelle Salt is a good example on which to make the measurements and test the reasoning that gives us the picture of crystals as constructed from building blocks. Looking down on one end of a crystal of Rochelle Salt, you get a plan view of it somewhat like Figure 51A. You can guess, from the symmetry of the plan

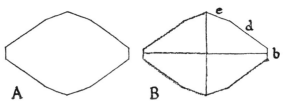

Fig. 51. AN "IDEAL" ROCHELLE SALT CRYSTAL (*A*) *has a symmetric shape, suggesting* (*B*) *that the angles between the faces e, d, and b are repeated around it.*

view, that you could divide the crystal into four parts, and find the same interfacial angles repeated in each part, as shown in Figure 51B.

This leaves you with only three angle measurements to make: the angles between *b* and *d*, *d* and *e*, and *b* and *e*. In fact, you need to make only *two* angle measurements: from any two of these angles you can calculate the third. The best angles to choose are those between *b* and *d*, and *b* and *e*.

Use the contact goniometer to measure and draw these two angles on a piece of paper, as shown in Figure 52,

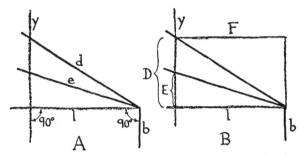

Fig. 52. A SIMPLE RELATION *between faces b, d, e on a Rochelle Salt crystal can be found by using the contact goniometer. A—Measure the angle between b and d, and the angle between b and e, carry them over to a piece of paper as in Figure 50, and then construct on the paper the lines l and y. B—Measuring the lengths D and E cut off on the line y, you will find that D is twice as long as E. In consequence a building block whose sides are in the same proportion as the proportion of D to F can be used to construct the crystal.*

and then add the lines *l* and *y*. Now examine the line *y;* in particular, measure with a ruler the lengths D and E on that line. You will find that length D is exactly two times length E.

The simplicity of that relation between the two angles

is proof that you can visualize the crystal as made of building blocks. Certainly you can choose a building block that will construct the angle between *l* and *d* in the way shown in Figure 53A. The simple relation be-

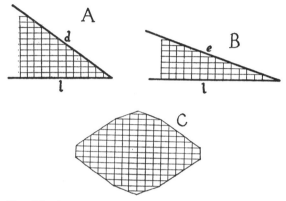

Fig. 53. A SINGLE SHAPE OF BUILDING BLOCK *constructs (A) the face d, and (B) the face e, and therefore (C) the whole crystal.*

tween D and E in Figure 52B makes it possible for the same building block to construct the angle between *l* and *e* in the way shown in Figure 53B. Thus you could construct the whole crystal from that one kind of blocks, as shown in Figure 53C. If lumps of sugar happened to come in the right proportions, you could make a model of the Rochelle Salt crystal in much the same way that the model of the alum crystal shown in Plate 5 was made of sugar cubes.

Over the years crystallographers have measured the angles between the faces on crystals of many different substances with great accuracy, and have never found one for which a suitable building block could not be chosen. Many crystals exhibit a great wealth of faces—a far greater variety than Rochelle Salt exhibits—but always a single shape of building block can account for all the faces. All the relations such as that of D to E in Figure 52B turn out to be simple: the ratios of the analogous lengths are simple fractions. Any face can be made by stepping back just a few blocks after piling up just a few rows.

Of course, crystallographers soon outgrew the contact goniometer. At best its accuracy is poor, and it is hopelessly awkward for a small crystal. They replaced it by the reflecting goniometer, and refined that instrument to give great accuracy.

You can easily make a remarkably accurate reflecting goniometer yourself by following the directions given in Figure 54A. It is especially easy to use on the seed crystals. For example, seed crystals of strontium formate dihydrate are good candidates for measurements from which you can determine the ratios of the sides of a rectangular building block, in the way shown in Figure 54B.

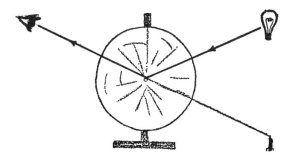

Fig. 54A. A "REFLECTING GONIOMETER," *easily made, measures angles between the faces on a crystal with quite high accuracy—as good as one-half degree when the faces are clean and bright. Moreover, since the instrument works best when the crystal is about the size of a seed, it can be used for crystals too small for the*

contact goniometer. First pivot a circular piece of thin cardboard through its center to a wooden upright with an upholstery nail. Then cement a similar piece of circular cardboard and a circular piece of polar coordinate paper together. Push a thumb tack (with a drop of glue on its under side) through the cardboard so that its point projects through the center of the coordinate paper. Cement the two pieces of cardboard together concentrically; the assembly then pivots on the upright, and the thumb tack revolves with it. Cement a small piece of paper to the upright and mark a vertical line on it, against which you can read the angle on the coordinate paper.

To use the instrument, mould a little conical piece of Plasticine against the coordinate paper and around the point of the thumb tack. Press the crystal into the Plasticine far enough to hold it. Set the instrument on a table, and place a source of light (such as a bridge lamp with the shade removed) two or three feet horizontally and a foot or two vertically from it. Turn the crystal-bearing assembly slowly on its pivot, and observe any flashes of light reflected from the crystal faces. Adjust the position of the crystal in the Plasticine until the faces whose angle you want to measure both give you flashes without your needing to move your head horizontally. Now move your head vertically until you line up your eye, the crystal, and some fixed horizontal line such as the edge of the table, or a ruler on the table, as the first diagram in this figure shows. Turn the assembly until you get a reflection, and read the angle. Turn the assembly again and read the angle for the second reflection. The difference of the two readings is the angle between the perpendiculars to the two reflecting faces. This angle always equals 180° minus the angle between the faces.

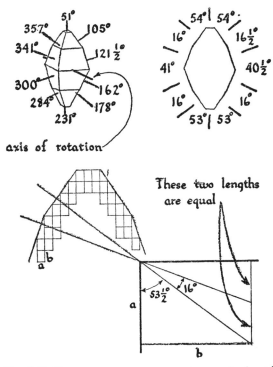

Fig. 54B. PLOTTING AND INTERPRETING *a set of readings taken with the instrument on a crystal of nickel sulfate hexahydrate. The differences between these readings are the measured angles between the perpendiculars to the faces. The averages of these differences for corresponding pairs of faces (16° and 53½°) are used to draw the two angles in the last part of this figure. As in the case of Rochelle Salt (Figure 53), the two angles turn out to be simply related, and to show that the crystal could be built out of building blocks with sides in the ratio of* a *to* b.

CHAPTER VII

The Symmetry of Crystals

A conspicuous feature of most crystals is their *symmetry,* and already symmetry has been helpful to you. In the discussion of the habit of alum, for example, it was possible to think of three kinds of faces—triangles, squares, and rectangles—repeated at various places around the crystal. And the symmetry of the Rochelle Salt crystal enabled you to reduce the number of angle measurements which you had to make to determine the proportions of its building block.

But the idea of symmetry is a much larger idea, useful in literature and design, in biology and mathematics, in language and logic. As is true of most ideas that are so large and so deep-seated that they are used almost unconsciously, symmetry often seems to elude a precise definition. Nevertheless, the kinds of symmetry which apply to crystals can be precisely defined. Indeed, a way can be found to specify exactly what kinds of symmetry an ideal crystal can have and what kinds it cannot have.

The study of the symmetries of crystals, combined with the study of how they can be constructed from building blocks, led crystallographers to a systematic scheme for classifying crystals into "crystal systems." A knowledge of the kind of symmetry possessed by a particular crystal turns out to be important also in studying the physical behavior of that crystal, as you will see in later chapters. And when you have learned how to look for symmetry, and how to describe the symmetry which you find, you will constantly be discovering it in the most unexpected places, and observing more and more the role it plays throughout nature and art.

The Idea of Symmetry

Certainly the most familiar example of a kind of symmetry is the symmetry of a human being. Viewed from outside, his left half is "like" his right half. Inside, of course, his organs lack that symmetry; he has a heart on the left side and none on the right side, for instance. Even outside, he may show slight differences between left and right—more muscular bulge on his right arm, or a parting in his hair on the left side of his head. But you have little trouble visualizing a symmetrical "ideal man," just as you can visualize an ideal crystal.

The letters of the alphabet will give you more illustrations of the presence or absence of symmetry. The letter A has the symmetry of the ideal man, but the letter F has none. The letters B, C, and D have a different symmetry: the top half of each is like its bottom half. And the letters H and I have both of these symmetries: their tops are like their bottoms, and their left sides are like their right sides.

Now look at the symmetries of some *words*. The word MOM has all the symmetry of the letter A. The word POP has not as much symmetry as MOM, but it still has a kind of symmetry: it is *spelled* symmetrically. In other words, if you ignore the unsymmetrical shape of the letter P, and pay attention only to what letter it is, you would say that POP is spelled symmetrically and MOP is not.

This kind of thinking about the symmetry of words illustrates a general problem in the idea of symmetry. When somebody says something is symmetrical, you have a right to ask him just what he means by "symmetrical." Consider, for example, how a logician may use the word, and how a mathematician may use it.

The statement, "John is a neighbor of Paul," might interest a logician in different ways. He might want to know whether the statement was true or false. But in either case the statement has one property which the logician immediately recognizes: it is "symmetrical." It is symmetrical in the sense that when you turn it around, to say, "Paul is a neighbor of John," you are saying the same thing that you said in the first place. On the other hand, "John is a son of Paul" is unsymmetrical. In the logician's uses, the word symmetrical has moved from the realm of physical things into the realm of thoughts.

The familiar process of adding two numbers together furnishes a simple example of symmetry in mathematics. "Two plus three" is mathematically symmetrical, because "three plus two" gives the same answer. On the other hand, "three minus two" is unsymmetrical.

Notice that the mathematician's use of symmetrical is really very similar to the logician's use. The logician says, "It makes no difference if I interchange John and Paul in the statement that John is a neighbor of Paul." The mathematician says, "It makes no difference if I interchange two and three in the expression 'two plus three.'" In fact, going back to words, letters, and the shape of a man, you see that you can give symmetry a similar description there also. You can say, "It makes no difference if I interchange the beginning and ending of the word 'Mom,' or the left and right sides of a man."

This is a clue to the ways of studying symmetry in

almost all the places where the idea is used, including crystallography. You ask, "Is there a way by which I can interchange parts of the ideal crystal and produce a result which looks just like the original crystal?" There may be no way, and the crystal then "lacks symmetry" or "is asymmetric." There may be several ways, and you can then describe the symmetry of the crystal by specifying what those ways are. All those several ways bundled together define the "symmetry class" to which the crystal belongs.

It turns out that there are just thirty-two possible classes of symmetry to which a crystal can belong, including the class of crystals having no symmetry at all. Each of these classes represents a bundle of "symmetry operations," which we define as a collection of things you can do, or things you can imagine doing, to an ideal crystal and leave it looking the same as it did before you performed the operation. Turning a crystal end for end is an example.

Symmetry Operations

In studying symmetry operations, it is best to begin with two-dimensional figures because they are simpler and easier to picture than three-dimensional figures. The letters of the alphabet again make a good starting place. Look first at the letter S.

If you turn an S around its center in the way shown in Figure 55 until its top has replaced its bottom, you reproduce exactly its original appearance in exactly its original place. Here then is a symmetry operation for that letter. Figure 56 shows that you could think of this operation as one in which you attach the center of the S to an axle perpendicular to the paper, and then turn the axle through one half of a full turn.

Fig. 55. TURNING THE LETTER S *about its center by one half of a full revolution leaves it looking as before.*

The center line of such an axle—the line about which the axle turns—is called the "axis of symmetry." For S it is called an axis of "two-fold" symmetry, because as you turn the axle through one full turn, you reproduce the appearance of the letter twice, once at a half turn and again at the full turn. Clearly an equilateral triangle has an axis of three-fold symmetry, and a square has

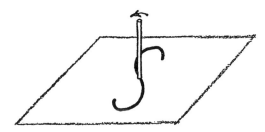

Fig. 56. AN AXIS OF TWO-FOLD SYMMETRY *lies along the center line of the axle used to turn the S.*

an axis of four-fold symmetry. The snowflake pictured in Plate 10 has an axis of six-fold symmetry perpendicular to it.

In short, the idea of an axis of two-fold symmetry provides a precise way of saying that the top and bottom of the letter S look alike. Now examine the letter K, whose top and bottom also look alike, but they look alike in a rather different way. Turning the K as you turned the S does not reproduce its appearance. But you can describe the symmetry of the letter K by another sort of operation, often called "reflection in a plane of symmetry." These two kinds of symmetry elements—axes of symmetry and planes of symmetry—describe the two fundamental kinds of symmetry operations in three dimensions also.

Before examining the definition of the second operation, notice a helpful way of picturing it in your mind. This way of looking at the operation makes clear why it is called "reflection in a plane of symmetry." If you draw the K on a piece of paper, and hold a mirror against the paper so that one edge lies along the horizontal center line of the K as in Figure 57, you will see an

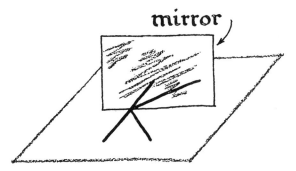

Fig. 57. REFLECTING THE LETTER K *in a mirror perpendicular to the paper and across the center of K makes K look the same as it looked before.*

entire K, of which one half comes from the K on the paper and the other half from its reflection in the mirror.

There is a very deep-seated distinction between axes of symmetry and planes of symmetry. Axes describe operations which you can actually perform physically. If your S is cut out of a piece of paper, you can stick a pin through its center and rotate it about the pin. Reflection in a plane of symmetry is an operation which you can perform only in imagination, or by an optical trick such as the use of a mirror. Figure 58 shows what

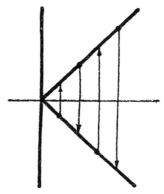

Fig. 58. A PLANE OF REFLECTION SYMMETRY *intersects the paper through the middle of K. Moving every point of the letter perpendicularly to the plane and through it to an equal distance on the other side, reproduces the letter.*

the operation is. You draw a horizontal line through the center of the K, and you imagine moving every point of the K perpendicularly across the line, stopping when the distance of the point from the line is the same as its original distance on the opposite side.

This is the kind of symmetry possessed by the outside of an "ideal man": a single plane of symmetry dividing him vertically in half. For example, when you reflect him in that plane, you transform his right foot into his left foot, and you transform his left foot into his right foot. In the unlikely event you ever found yourself in the ridiculous plight of having two left shoes, you would quickly learn that there is no way of rotating one of them into a right-foot shoe. You would have to perform the imaginary operation of *reflecting* one of them into a right-foot shoe.

The fact that rotation is a "performable operation," whereas reflection is a "nonperformable operation," has another familiar consequence. Usually a person sees his own face only after it has been reflected by mirrors or shop windows. When he sees a photograph of himself, he may disagree with friends who say, "That looks just like you." For nobody has an "ideal face": there are subtle differences between anybody's left side and his right side. His friends are used to seeing those differences in their right places, while he is used to seeing those differences on the wrong sides, and no amount of twisting and turning will ever make his friends' image of him coincide with his image of himself.

This experience suggests that you look a little more generally at these operations, which are capable of being symmetry operations if they are performed in the right ways on the right objects, and notice what they do to an object when they are *not* symmetry operations for the object. Take a human hand, which has no symmetry, as an object to operate on; notice in Figure 59A what happens when you reflect it in any plane. You transform it from a right hand into a left hand. Now reflect that left hand in a second plane, at right angles to the first, as shown in Figure 59B. The operation transforms it back into a right hand, but upside down. You could have produced that last hand from the first hand by a single operation: rotating the first hand one-half turn about an axis

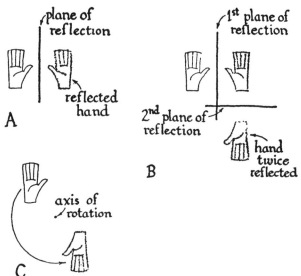

Fig. 59. REFLECTING AN OBJECT TWICE *is equivalent to turning it. A reflection (A) is called a "nonperformable operation," because it accomplishes the impossible feat of turning a right hand into a left hand, for example. But a second reflection (B) turns the left hand back into a right hand. This succession of two "nonperformable" operations is equivalent (C) to the single performable operation of rotation about the line of intersection of the two planes.*

lying along the intersection of the two planes, as shown in Figure 59C.

This is an example of a rule applying to all these operations. A succession of two nonperformable operations is equivalent to a single performable operation. If a person wants to see himself as others see him, he need not pay a photographer to take a picture of him. He can look at himself in two mirrors. The first turns him the wrong way round, and the second turns him back again.

There is one other nonperformable symmetry operation important in crystallography. The operation is not really independent of the operations of rotation and reflection; but it is convenient to think of it separately because there are crystals whose symmetry consists of this operation alone. The letter S can be used to illustrate it.

You have already noticed that turning the S a half turn about its center reproduces the S. Another way you could reproduce the S would be to move all its points along lines toward the center, through the center, and out from the center again to an equal distance on the other side. Figure 60 shows the operation. It is called

Fig. 60. INVERTING AN S *through its center leaves its appearance unaltered. By inversion every point is moved through the center to an equal distance on the other side of the center.*

"inversion through a center," and the center is called a "center of inversion," or more often a "center of symmetry." The example of S may make you feel that there is no reason to distinguish this operation from rotation by a half turn. All the points move to the same places whether you turn them around the center by a half turn or invert them through the center. As a matter of fact, there is no reason to distinguish the two operations on figures in two dimensions. But the operations have distinguishable effects in three dimensions, as you will see soon when you look at the symmetries of crystals.

Before turning to those symmetries, you can get valuable practice in finding the symmetry operations appropriate to objects by examining the rest of the letters of the alphabet and placing them in symmetry classes.

PROBLEM 6

Each letter of the alphabet belongs to one of the following five classes of symmetry. List the letters in their appropriate classes.

Class	*Example*
1. No symmetry	P
2. Two-fold axis	S
3. Vertical plane	A
4. Horizontal plane	K
5. All of 2, 3, and 4	H

Symmetry in Three Dimensions

Already you have examined the symmetry of a three-dimensional object, the "ideal man." His symmetry is low: there is only one operation that would leave his external appearance unaltered—reflection in a plane. There are crystals with the same low symmetry. Indeed there are crystals that have the same symmetry as the insides of man: none at all.

On the other hand, there are crystals whose symmetry is very high—for example, alum and common salt. As you noticed in Plate 4, common salt habitually forms little cubes. You can determine the symmetry of common salt by finding the symmetry operations appropriate to an ideal cube.

Unless your ability to visualize figures in three dimensions is unusually great, the best way to study the symmetry of any polyhedron is to hold an example of it in your hands and look at it from many different points

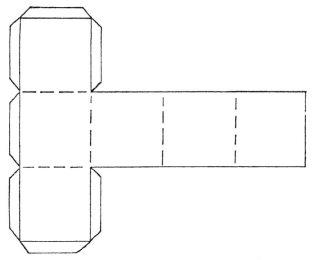

Fig. 61. MODELS OF POLYHEDRA *are made by cutting, folding and gluing a piece of heavy paper or light cardboard. The pattern shown makes a cube. Score the paper with a dull knife, to facilitate folding, and leave tabs for gluing. Rubber cement and contact cement are good adhesives. The perforated tape used to seal the cracks in plaster-board walls is an excellent material for polyhedra models. It is punched with holes which facilitate accurate bending at several angles. Some polyhedra are most easily made from several pieces of cardboard or tape, held together by adhesive cellophane tape.*

of view. Figure 61 shows an easy way to make models of polyhedra.

It is helpful to systematize your search for symmetry operations in the following way. Look first for axes of symmetry, and when you have found them all, begin looking for planes of symmetry. In searching out axes, look first for axes passing through *faces,* then for axes passing through *edges,* and finally for axes through *corners.*

To follow this program with a paper or cardboard cube, you pass a thin stick or a piece of stiff wire through the centers of two opposite faces, as shown in Figure 62. If you rotate the cube one quarter of a full

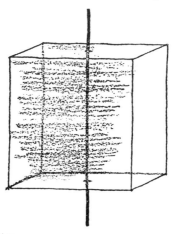

Fig. 62. AN AXIS OF FOUR-FOLD SYMMETRY *passes through the centers of opposite faces of a cube. The cube has three such axes.*

turn around the wire, the cube looks the same as it did before you rotated it. In one full turn, the cube looks the same four times, once for each quarter turn, no matter where you start. Hence, the wire is marking an axis of four-fold symmetry.

A cube has six faces arranged in three pairs. The members of each pair are opposite and parallel, and you have found a four-fold axis through one of the pairs. Clearly, there is a similar four-fold axis through each of the other two pairs. Thus you have found three four-fold axes, perpendicular to one another.

A cube has twelve edges arranged in pairs whose members are opposite and parallel. Passing the wire through the centers of two opposite edges, and turning the cube, you find an axis of two-fold symmetry, as Figure 63 shows. Since there are six similar pairs of edges,

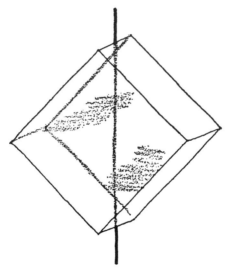

Fig. 63. An axis of two-fold symmetry *passes through the centers of opposite edges of a cube. The cube has six such axes.*

there are six axes of two-fold symmetry.

The eight corners of a cube are again associated in pairs—four pairs of opposite corners. A wire through any pair of opposite corners, as in Figure 64, lies along

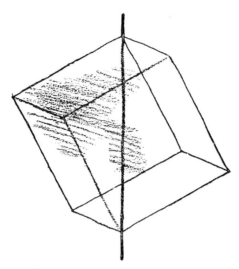

Fig. 64. AN AXIS OF THREE-FOLD SYMMETRY *passes through opposite corners of a cube. The cube has four such axes.*

an axis of three-fold symmetry, and thus the cube has four such axes. The fact that a cube has any *three-fold* symmetry surprises most persons. Look carefully down the axis, as in Figure 65, to convince yourself. The oc-

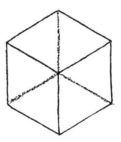

Fig. 65. THE THREE-FOLD SYMMETRY OF A CUBE *is conspicuous when the cube is viewed along an axis of three-fold symmetry, a line between opposite corners.*

tahedron faces on alum crystals are perpendicular to
these three-fold axes, and you have already noticed that
a crystal of alum can consequently look triangular.

There is no way to divide the search for planes of
symmetry quite as systematically as the search for axes.
But it is true for some figures—the cube is one—that there
are planes of symmetry which *cut through edges* but
contain no edges, and there are also planes which *contain edges* but *cut through no edges.*

In the cube the planes of the first sort are like that
shown in Figure 66. There are three such planes of sym-

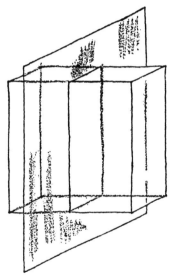

Fig. 66. A PLANE OF REFLECTION SYMMETRY. *This
one is parallel to a pair of faces. The three planes of
this type are often called the "cubic planes."*

metry, each parallel to one of the pairs of opposite faces
and midway between them. Each plane divides the cube
in half in such a way that each half is the reflection of

the other in that plane. One of the planes of the second
sort is shown in Figure 67; it contains a pair of opposite
edges. Since there are six pairs of opposite edges, there
are six planes of this sort.

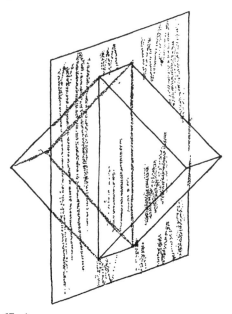

Fig. 67. ANOTHER PLANE OF REFLECTION SYMMETRY.
*This one contains a pair of opposite edges. The six
planes of this type are often called the "dodecahedral
planes," because they are parallel to the six pairs of
faces of a rhombic dodecahedron (Figure 41).*

Adding up all the symmetry operations you have
found for the cube, you can now define its symmetry:
it consists of three 4-fold axes, four 3-fold axes, six
2-fold axes, and nine planes.

Often a quick way to find the symmetry of a figure
less symmetrical than a cube is to imagine how you
might make the figure by mutilating a cube. Then you

can look at each symmetry operation for the cube, deciding whether the mutilation would remove that particular symmetry operation. The symmetry of the new figure consists of the symmetry operations still remaining after you have completed this process of elimination.

For example, if you stretch a cube along one of its four-fold axes of symmetry, you produce the polyhedron shown in Figure 68. In order to examine the remaining

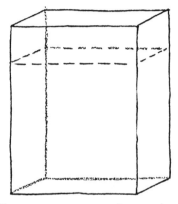

Fig. 68. STRETCHING A CUBE *along a four-fold axis destroys some of its symmetry.*

symmetry, think of the polyhedron as if it had been produced from the cube by pulling two opposite faces away from each other. Then one of the original four-fold axes is undisturbed, but the other two are reduced in status to two-fold axes. The original three-fold axes are all destroyed. Of the six original two-fold axes, only two remain. The three planes parallel to faces are undisturbed, but four of the other six planes disappear. You are left with one 4-fold axis, four 2-fold axes, and five planes. Figure 69 shows a crystal shape with the same symmetry.

Another interesting shape obtainable by distorting a

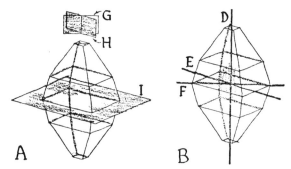

Fig. 69. A CRYSTAL SHAPE *with the symmetry of the stretched cube of Figure 68. A—There are two planes of symmetry of the type G, two planes of symmetry of the type H, and the plane of symmetry I. B—There are two 2-fold axes of the type E, two 2-fold axes of the type F, and the 4-fold axis D.*

cube is the rhombohedron. A rhombohedron is the figure you can imagine producing if you squashed a cube along one of the lines from corner to opposite corner, as Figure 70 shows. Just as common salt has the symmetry of a cube, calcite has the symmetry of a rhombohedron.

Distorting the cube into a rhombohedron destroys the

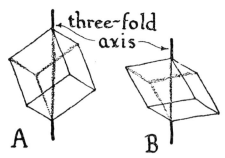

Fig. 70. SQUASHING A CUBE (*A*) *along a three-fold axis produces a rhombohedron* (*B*).

three 4-fold axes completely, and leaves only one of the
3-fold axes—the axis from corner to opposite corner
along which you squashed the cube. Of the six 2-fold
axes, only three remain. They are the three 2-fold axes
perpendicular to the 3-fold axis that you singled out,
and they continue to be perpendicular to it in the rhombohedron. Three of the nine planes of reflection symmetry survive—the three containing the 3-fold axis. In
short, the symmetry of the rhombohedron, and therefore of calcite, can be described by one 3-fold axis,
three 2-fold axes, and three planes.

In addition to rotation about an axis and reflection
in a plane, the operation of "inversion through a center"
is a symmetry operation frequently encountered in crystals. Studying its application to the letter S, you noticed
that in two dimensions inversion was not distinguishable
from rotation through a half turn. In three dimensions,
however, it becomes distinguishable, as you see in Figure 71.

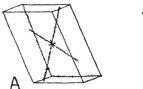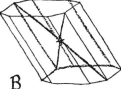

Fig. 71. INVERSION THROUGH A CENTER *is the only
symmetry operation for the parallelepiped (A) and the
"ideal" crystal of blue vitriol (B).*

Like the letter S, an ideal crystal shape is said to have
a "center of inversion" or a "center of symmetry" if,
when an imaginary line is passed from any point on its
surface through its center, it meets a corresponding
point on the opposite side of the crystal. The distance
between those two points and the center must be the

same, for all possible lines through the center. The shape in Figure 71A is the "general parallelepiped." It has three pairs of opposite parallel faces, like the cube, but each pair is inclined to the other pairs. Each face is a "general parallelogram," without right angles and with no special relation between the lengths of its two pairs of opposite sides. The parallelepiped has *only* a center of symmetry and no other symmetry operation.

Even though inversion through a center in three dimensions is different from rotation through a half turn, the operation is still not a completely new one. In fact, it is equivalent to rotation through a half turn followed by reflection in a plane perpendicular to the axis of rotation. Figure 72 shows that these two operations, ap-

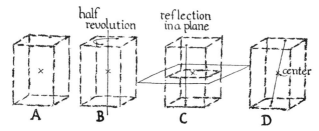

Fig. 72. INVERSION THROUGH A CENTER, *as can be seen by operating on a point (A), is equivalent to a succession of two operations: rotation through a half-turn about an axis (B) followed by reflection through a plane perpendicular to that axis (C). The equivalent center of inversion (D) is at the intersection of the axis and the plane.*

plied in succession to any point, move the point to the same place to which inversion would move it. The center of inversion is the point of intersection of the axis of rotation and the plane of reflection. In this equivalence, notice that you could choose any direction through the center of inversion as the axis of rotation, so long as you

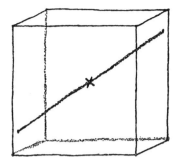

Fig. 73. INVERSION THROUGH A CENTER *is one of the symmetry operations of a cube.*

also choose the plane perpendicular to that axis as the plane of reflection. In two dimensions a center of inversion is equivalent to a two-fold axis because the points to be moved all lie in the plane of reflection, and therefore reflection does not move them.

Figure 71B shows a crystalline shape—that of the familiar blue vitriol, copper sulfate pentahydrate—which, like the parallelepiped, has only a center of symmetry. As Figure 73 shows, a cube has a center of symmetry. On the other hand a tetrahedron has no center of symmetry.

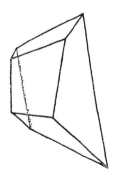

Fig. 74. A POLYHEDRON WITHOUT SYMMETRY.

Figure 74 pictures a polyhedron having no symmetry whatever. It too is a possible crystal shape; the fact that the atoms in a crystal are arranged in repetitive order does not imply that their arrangement must exhibit symmetry.

The Six Crystal Systems

Crystals usually do have symmetry, however, and they can always be imagined to be made of little building blocks. In combination these two characteristics have interesting consequences. The building block is somewhat fictitious, of course. Nobody supposes that the ultimate fine structure of crystals looks like a collection of brick-shaped particles. The assembly of identical building blocks is merely a way of portraying the crystal's repetitive orderliness. Each building block represents a grouping of atoms or molecules which is regularly repeated throughout the crystal.

Since the whole crystal can be built of identical building blocks, just by repeating enough of them, any one of those building blocks contains all there is to know about the arrangement of atoms in the entire crystal. For example, if the length of the building block is much greater than its width, then the distance at which the atomic arrangement repeats itself is much greater along one direction than along another. But the block cannot yield all its atomic information in its shape alone. Most of the information is locked inside the block, so to speak, and the block yields that information only when the arrangement of atoms within it has been discovered.

Nevertheless, a study of the possible shapes of building blocks, even without knowing the details of their contents, sheds additional light on the symmetries of crystals. In particular, the study makes clear that there are limitations to the kinds of symmetries crystals can have. You might ask, for example, "Can a crystal have a five-fold axis of symmetry?" To answer this question, you can reason that if a crystal has such an axis, then it must be possible to choose a building block whose shape also has such an axis. The building blocks might be the five-sided cylinders shown in Figure 75, whose ends are regular pentagons.

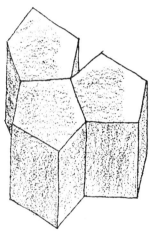

Fig. 75. FIVE-SIDED BUILDING BLOCKS, *whose cross sections are regular pentagons, do not fit together to fill space, and are therefore not admissible building blocks.*

But such a block is inadmissible. Building blocks for crystals must fit together tightly to fill space, without leaving gaps between them, and regular pentagons do not fill a plane. There is no permissible form with an axis of five-fold symmetry, and therefore no crystal can have such an axis.

It is interesting to reflect upon the difference between crystals and living organisms in this respect. Starfish and buttercups, for example, have axes of five-fold symmetry; but crystals cannot. Examining the other sorts of axes of symmetry in the same way, you will reach the conclusion that the only kinds of axes a crystal can have are two-fold, three-fold, four-fold, and six-fold.

The classification of crystals into crystal systems is made on the basis of the kind of building blocks appropriate for the crystal. It turns out that there are six appropriate sorts of blocks, and to them the six crystal systems correspond in the way shown in Plate 28.

Do not make the mistake of thinking that a crystal necessarily has all the symmetry shown by the shape of its most appropriate building block. The least symmetrical of all the blocks is the triclinic block shown in Plate 28. It has the shape of a parallelepiped, which has a center of symmetry, as you have already seen. But there are crystals with no symmetry whatever, and such crystals must therefore have a lower symmetry than their building block.

Sodium bromate, which takes the shape of a regular tetrahedron, is an example of a crystal whose symmetry is clearly lower than the symmetry of its building block. A regular tetrahedron can be built up of cubes, in much the same way our model of alum (Plate 5) was made of cubes, because the tetrahedron can be made by extending half of the faces of the octahedron until they

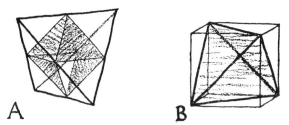

A   B

Fig. 76. A TETRAHEDRON RESULTS *from (A) extending half the faces of an octahedron. But the best way of drawing a regular tetrahedron is by connecting half of the corners of a cube (B).*

meet (Figure 76). For this reason, sodium bromate belongs to the cubic crystal system. But the symmetry of a tetrahedron is much lower than the symmetry of a cube.

You will see more clearly why crystals can have lower symmetries than their building blocks when you look in the next chapter at some of the arrangements which atoms adopt in crystals. And in a later chapter you will take up the problem of assigning a crystal to its proper symmetry bundle—its "crystal class."

CHAPTER VIII

The Arrangements of Atoms
in Crystals

As the discussion in the last two chapters proceeded, it moved away from those real, solid objects, crystals. First they became "ideal crystals," and then empty "polyhedra" made of fictitious "building blocks." This process of abstraction is typical of much scientific thinking. Scientists believe that by thinking of an ideal and apparently empty world they learn about a solid world outside their minds, where things really happen. But they know that their studies would not have relevance to reality if their ideal world were wholly fanciful; it must make a clear connection with the real world.

In order to begin re-establishing the connection between the real and ideal worlds of crystallography, recall that crystals are built not of building blocks but of atoms. The building blocks represent the atoms by showing something of how they are arranged in repetitive orderliness, and something of the symmetry which that orderliness exhibits.

But if you regard the building blocks as empty containers, they do not seem to portray all you might wish. For example, their shapes are often more symmetrical than the crystals which they build. Nevertheless, since a repetition of identical blocks constructs an entire crystal, you have reason to believe that a single block locks within it all the information you seek. The difference in the amounts of information conveyed by an empty

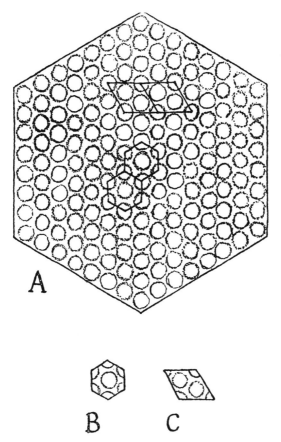

Fig. 77. THE CLOSE-PACKED ARRANGEMENT *in two dimensions* (*A*) *can be constructed of the hexagonal building block* (*B*) *or the rhombic building block* (*C*). *Each building block accounts for three atoms.*

building block and a block furnished with its atoms will emerge as you examine in this chapter how a building block represents atoms.

Two-dimensional Building Blocks and Atoms

Again, in this examination, it is easier to begin in two dimensions than in three. Figure 77A shows a simple repetitive arrangement of "two-dimensional atoms." It is the so-called "close packed" arrangement, in which circular atoms would occupy a minimum of space. Each atom is immediately surrounded by six others.

By blowing air at a uniform pressure through a capillary tube dipped beneath the surface of a pool of soapy water, you can, in fact, "grow a two-dimensional crystal," and watch the "atoms" take this arrangement. The little bubbles rise to the surface, attract one another slightly, and come together to form "bubble rafts," such as that pictured in Plate 29, each bubble touching six others.

You can construct a building block for this arrangement by drawing lines between the centers of some of the atoms, to form a regular hexagon, as shown in Figures 77A and B. The hexagon is a true building block for the arrangement. If you repeat it, carrying its proper proportion of each atom with it, you can construct the whole crystal; the repeated blocks all fit together in such a way as to leave no space unaccounted for.

This example shows that a building block is not necessarily a container literally "containing" its atoms. You can think of this building block as *accounting for* three atoms, but not as containing three entire atoms within its boundaries. It contains one third of each of its six corner atoms and all its central atom (Figure 77B).

A building block serving this arrangement just as well would be a rhombus whose outlines are obtained by connecting the centers of four of the hexagonal blocks. Again the rhombus accounts for three atoms: one sixth of two corner atoms, one third of two other corner atoms, and the two atoms wholly included, as Figures 77A and C show.

But to speak of one sixth of an atom or one third of an atom as included in the block at a corner is really to speak too literally. In the first place, such language assumes that the atom has a circular shape—a risky assumption. In the second place, in order to count up the number of atoms represented by a building block, it is not necessary to know the exact fraction of an atom included at a corner. It is only necessary to count the number of occupied corners.

The reason for the last statement becomes clear as soon as you recall the meaning of a building block: a unit which can be repeated, without turning it, to build

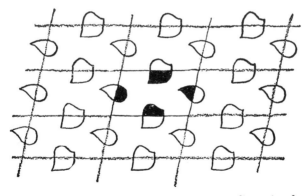

Fig. 78. COUNTING THE ATOMS *in a two-dimensional block. An atom on a side of a block counts one half in calculating the contents of the block. If the block is a true building block any atom on one side of the block must have a companion on the opposite side, and the two atoms contribute a count of* one *to the contents of the block.*

up the whole crystal. From that meaning it follows that, if there is an atom at one corner of a four-cornered block, there is an atom at all four corners of the block. It also follows that each corner of a four-cornered block is shared by three other blocks. Therefore, in a two-dimensional crystal any corner atom belongs to four blocks, each carrying its share of the atom. The four shares together represent the entire atom, whatever its shape. You can say that each corner atom counts a quarter for any four-cornered block, and that there are four of those quarters for any one block.

Similar reasoning gives a count of one half to each atom on a *side* of a two-dimensional block. Figure 78 illustrates how this method of counting will always yield a whole number of atoms for any true building block. In three dimensions similar reasoning justifies a count of one eighth for each atom at a corner, one quarter for each atom on an edge, and one half for each atom on a face.

Now contrast the hexagon of Figure 77B, furnished with its three atoms, and an empty hexagon. The same hexagon is just as good a building block for the different arrangement of atoms shown in Figure 79. There the block has the same shape and size, but each block accounts for only two atoms instead of three. Hence the empty hexagon is not enough to show the atomic arrangement in a two-dimensional crystal. The hexagon must be furnished with its atoms.

It is worth while stopping to compare these two arrangements. For both arrangements—that in Figure 77 and that in Figure 79—all the atoms are alike. Of course, the substances which crystallize in these arrangements would probably be two different substances, because any one substance usually crystallizes in only one arrangement. But each substance contains just one kind of atom,

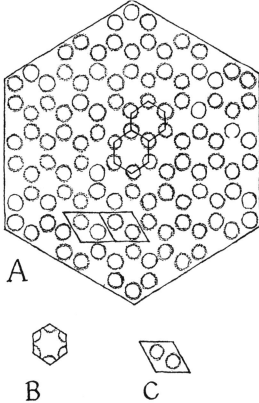

Fig. 79. THE "OPEN" HEXAGONAL ARRANGEMENT *in two dimensions* (A) *can be constructed of either a hexagonal building block* (B) *or a rhombic building block* (C) *having the same shapes as those for the close-packed arrangement in Figure 77 but accounting for only two atoms instead of three.*

much as a crystal of copper or a crystal of zinc contains only atoms of copper or atoms of zinc.

Certainly, the most conspicuous contrast between the arrangements is the difference in how closely the atoms are packed together. There are unoccupied "holes" in one arrangement; in the other they are filled. One result of these holes is that each atom in Figure 77 has six near neighbors and each atom in Figure 79 only three.

From that difference you can infer a difference in the behavior of the two kinds of atoms. In the close-packed arrangement the atoms are behaving as if they were indeed little circular discs. The discs seem to be attracting one another and trying to get as close together as possible. They give the impression that each atom would like to have even more than six nearest neighbors if there were room for more atoms to crowd around it.

In the more open arrangement each atom is behaving as if it were satisfied when it has attracted three others. There is room for more than three, but the atom rejects the rest. Each atom gives the impression that it is putting out tentacles in three directions, engaging the tentacles of three other similar atoms. Its behavior strongly suggests the behavior of atoms in molecules, where each atom makes a strong bond with just a few others.

Real crystals provide instances of both kinds of behavior, and typical atomic arrangements are pictured later in this chapter. Crystals of copper and of zinc, for example, have arrangements in which the atoms seem to be behaving like little spheres packed as closely as possible. Each atom has twelve nearest neighbors. In crystals of silicon, each atom has only four nearest neighbors. Iron is intermediate between these two extremes; each atom in a crystal of iron has eight nearest neighbors.

In the close-packed arrangement of two-dimensional atoms (Figure 77), you may have noticed already the possibility of choosing a smaller building block—a block (Figure 80) containing only one atom instead of three. It is not surprising that this can be done when the atoms are all of the same kind and all behaving in the same

Fig. 80. A ONE-ATOM BUILDING BLOCK *can construct the close-packed arrangement of Figure 77.*

way. It is more surprising that you cannot choose a building block containing only one atom when the atoms have the open arrangement of Figure 79.

The reason why a one-atom block is impossible in the second case becomes clear when you notice that the atoms are playing two slightly different roles in this arrangement. The nearest neighbors of half of the atoms are located as Figure 81A shows. Around the other half

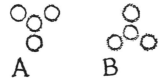

Fig. 81. TWO DIFFERENT ORIENTATIONS. *The atoms in the hexagonal arrangement of Figure 79 can be divided into two classes which play different roles in the arrangement. The center atoms at A and B are examples. Hence the smallest building block must contain two atoms, one of each class.*

the location of neighbors is as shown in Figure 81B. The atoms are all of the same kind, and connected to their nearest neighbors in the same way, but those neighbors are differently disposed in the arrangement. Hence, the smallest building blocks for the arrangement are two-atom blocks such as those shown in Figures 79B and 79C. In real crystals this difference becomes important.

A one-atom building block can be chosen for copper, as you will see soon, but the smallest possible building block for silicon contains two atoms.

After dwelling on the differences between the close-packed and open arrangements in two dimensions, you should notice an important similarity between them. It goes beyond the fact that you can choose identical hexagons for their empty building blocks. Even when they are furnished with their atoms, so that the two hexagonal blocks differ, they still have the same symmetry as the empty hexagon: a six-fold axis, six planes of reflection, and a center of symmetry, as you see in Figures 82A, B, and C.

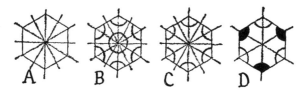

Fig. 82. HEXAGONAL BUILDING BLOCKS *may not have the symmetry of the empty hexagon (A) when they are furnished with their atoms. The blocks for the close-packed structure (B) and the open structure (C) retain full hexagonal symmetry. But the block (D) in which the atoms are of two different sorts, has a lower symmetry than the empty hexagon. The six-fold axis is reduced to a three-fold axis, and the center of inversion and three of the six planes are lost.*

For the atomic arrangement shown in Figure 83, the same hexagon will serve again as the empty building block. This fictitious two-dimensional crystal is made of two kinds of atoms. Thinking of the white circles as representing positive ions of some atomic species, and the black circles as negative ions of a different atomic species, you can see that the arrangement is a reasonable

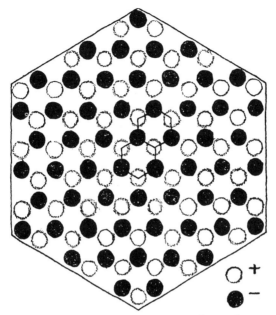

Fig. 83. AN IONIC ARRANGEMENT, *similar to the open hexagonal arrangement in Figure 79, with atoms of two kinds. It is a reasonable structure for a "two-dimensional ionic crystal," in which the positive white ions repel one another and attract the negative black ions.*

one for a "two-dimensional ionic crystal." Just as in sodium chloride, discussed in Chapter I, each ion has collected ions of the opposite sign around it, and has pushed ions of the same sign away.

In this case the building block loses some of its symmetry when it is furnished with its atoms. The furnished block, shown in Figure 82D, has lost its center of symmetry, and three of its reflection planes; it retains only three planes of symmetry, and a three-fold axis in place of a six-fold axis. Here is an example in two dimensions of how a crystal can have a lower symmetry than its empty building block. Sodium bromate, whose most appropriate building block has the shape of a cube, is an example of a real crystal with symmetry lower than that of its empty building block.

Since the symmetry of one kind of crystal may be the same as that of its building block, and the symmetry of another kind may be lower, the six crystal systems form only a beginning of a scheme for classifying the symmetries of crystals. When further on you will examine the possible symmetries of crystals in a more detailed way, you will find that there are thirty-two classes of crystalline symmetry. Each class belongs to one or another of the six crystal systems. In each system there is one class with the same symmetry as the empty building block for that class. The other classes have less symmetry than the blocks. A crystal will belong to one of those less symmetrical classes when the building block furnished with atoms has a lower symmetry than the empty block.

The Structures of Real Crystals

Until about fifty years ago there was no experimental way to find the atomic arrangement in a crystal—the way in which the building block is furnished. It was only possible to get two kinds of information about the arrangement. The shape of the building block could be determined by making the measurements of interfacial angles, and the symmetry of the atomic arrangement could often be learned from the habit of the crystal and other properties.

In 1912 three German physicists, Max von Laue, W. Friedrich, and P. Knipping, performed a very important experiment. They showed that a crystal will scatter a beam of X rays into a large number of separate beams, which come out of the crystal in definite directions. Another German, Wilhelm K. Roentgen, had discovered X rays only seventeen years earlier. Nobody was certain what the rays were, but most physicists supposed that they were a form of light with very short wave lengths. Coupling that guess with the long-standing guess that the atoms in a crystal had a regularly repeated arrangement, the three physicists in their experiment confirmed both at once. The spots you see in Plate 30 were made by the beams of X rays scattered onto a photographic plate by a crystal of quartz.

Since that time the work of many investigators has perfected methods of scattering X-ray beams from crystals and interpreting the angles at which the scattered beams emerge. The investigations yield information on how the atoms are arranged and even how far apart the atoms are. Today the "crystal structures" of many materials have been worked out.

But there are still limitations to the information X rays can yield. Except in rare cases, X rays can only propose a few alternative structures and cannot assert which of those few is the most probable. In order to narrow the final choice down to a single structure, it is necessary to use other information. The habit of the crystal, its optical properties, the "etch figures," which we will consider later—any of these properties may be useful.

For complicated crystals the final choice must depend on an "educated guess," based on a knowledge of how atoms combine into molecules in chemical compounds. In fact, the relationship of crystallography and chemistry has a long history and continues to work both ways. X-ray determinations of crystal structure often help the chemist to discover the manner in which the atoms are interconnected in complicated organic molecules—for example, in the molecule of penicillin.

After the structure of a crystal has been determined,

one of the best ways to make a picture of it is to show
how the atoms are arranged in just one of its building
blocks. After all, no more is necessary, for the whole
crystal is built up by simply repeating the block. Today
the building block furnished with its atoms is usually
called a "unit cell."

It is a little more difficult to make a realistic picture
of an arrangement of atoms in three dimensions than in
two dimensions. If you make pictures in which the atoms have their proper sizes for the scale of the picture,
the atoms hide one another, as you see in Plate 31. For
this reason drawings or models of the arrangement of
atoms in a crystal usually have little spheres located
where the centers of the atoms would be. The spheres
are connected by enough lines, as in Figure 84, to guide
your eyes so that you can see the arrangement clearly.

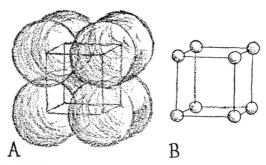

A   B

Fig. 84. INDICATING POSITIONS OF ATOMS *by small
spheres gives a clearer picture (B) of the array than a
diagram (A) in which the "atoms" are shown in their
proper relative size.*

Looking at diagrams and models such as these, you
must be careful to avoid two tempting misconceptions.
Because the spheres are widely separated, you may forget that the atoms themselves are really in contact with
their neighbors. You may also feel that the lines in the

drawings, and the sticks in the models, represent bonds between the atoms, but they are only visual aids in the drawings, and mechanical members necessary to the construction of the models. Usually some of the guide lines outline the unit cell, but others do not.

According to this custom, the crystal structure of sodium chloride—the arrangement of ions which you have already looked at in Figure 13—would look like Figure 85A. The structure shown is that of one unit cell; the

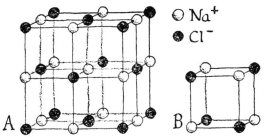

Fig. 85. A UNIT CELL OF SODIUM CHLORIDE (A) containing four ions of each sort. The cubic fragment (B) is not a unit cell because it does not build the structure when properly repeated.

method of counting already described yields its "contents": four sodium ions and four chloride ions.

You might be tempted to assert that you could choose, as a smaller unit cell, the portion shown in Figure 85B. But this is not a true unit cell because it is not a true building block: repeating it so as to leave no empty spaces, you would have to place sodium ions in the same places as chloride ions. Another way to see that it cannot be a true unit cell is to notice that it would account for only half of a sodium ion and half of a chloride ion. A true building block will always account for whole numbers of atoms.

Nevertheless, there is a smaller unit cell for the structure of sodium chloride—a cell which accounts for just one ion of each sort. You will see how to find this smallest cell soon, after you have studied some ways of analyzing structures.

Close-packed Structures

Among the simplest structures which crystals adopt are those called "close-packed." These arrangements would be expected in a crystal made of a single kind of atoms, when the atoms are behaving like little spheres which attract their fellows in all directions. Each atom would collect around it as many of its fellows as it could, as close to it as possible. Figure 77 has shown what would happen if the crystal were two-dimensional and the atoms were little circular discs, and Plate 29 has shown the corresponding behavior of soap bubbles collecting on the surface of a pan of soapy water. Marbles, brought together in a single layer on a plane surface, can be assembled into the same hexagonal arrangement, in which each marble is immediately surrounded by six others.

When you extend the arrangement of marbles into three dimensions, you meet some interesting complications. A good way to study them is to pile up successive layers of marbles so that each layer is close-packed. The pile can be held together if you place the first layer in a hexagonal array of holes, like that on a Chinese checkerboard, drilled in a piece of plywood.

The second layer fits neatly over the first layer. Each marble in the second layer is supported by three marbles in the first layer, as Plate 32 shows; and, of course, a third layer can be supported by the second layer in the same fashion. After the addition of the third layer, each marble in the second layer is completely embedded, and makes contact with twelve others—six in its own layer, three in the first layer, and three in the third layer.

But if you look straight down on the marbles as you add the third layer, you will see the beginning of the complications offered by close packing in three dimensions. You have two alternative ways of adding the third layer. You can place the marbles either directly over marbles in the first layer, or over open spaces in the first layer. Either way of arranging the third layer looks the same to the second layer. But the two ways look different to the first layer, as Plate 33 shows. Clearly, the two ways of arranging the marbles are equally close-packed, and either way puts each marble in the second layer in contact with twelve others.

Now you could continue to repeat either of these arrangements to form orderly structures. If you choose to put marbles in the third layer over those in the first layer, those in the fourth layer over those in the second layer, and so on, you will build up one kind of orderly close-packed structure. If, instead, you put marbles in the third layer over open spaces in the first layer, marbles in the fourth layer over open spaces in the second layer, and so on, you will build up another kind.

A convenient way to begin analyzing structures of these kinds is to distinguish by the letters A, B, C the possible positions of the marbles. Call the positions of the marbles in the first layer "A-positions," and those in the second layer, which are over open spaces in the first layer, "B-positions." Then if the marbles in the third layer fall over those in the first layer, they too occupy A-positions. If the marbles in the third layer fall over open spaces in the first layer, they occupy the only remaining new sorts of positions, which you can call "C-positions."

These two orderly ways of stacking the marbles can now be described with the letters. The first manner of

stacking is a repetition of the sequence A B A B A B
. . . The second manner is a repetition of the sequence
A B C A B C . . .

This way of analyzing structures suggests at once that
there must be an unlimited number of possible orderly
close-packed arrangements. The regular repetition of
A B provides an orderly way of stacking the layers which
repeats itself every two layers. In the regular repetition
of A B C, the stacking order repeats itself every three
layers. You can easily invent structures with longer repeating cycles, such as A B A C (a cycle of four) or of
A B A B C (a cycle of five). Any such repeated arrangement would be orderly, and therefore crystalline.
Indeed, a disorderly choice of the stacking order would
still produce a close-packed structure, although no
longer a crystalline one.

It is interesting to notice that the two close-packed
structures first described occur very much more widely
than any others. Of course, you would expect to find
such structures only in crystals made of a single kind
of atoms. And those atoms must not combine with one
another in strong chemical bonds to form molecules in
which the atoms are tightly connected in pairs or rings
or chains. Two classes of substances fill these require-
ments—the metals and the "rare gases" such as argon
and neon. In fact, both classes furnish many examples
of the two structures—the metals at ordinary temperatures and the rare gases when they are solidified at very
low temperatures.

No doubt the reason why the close-packed structures
with a longer repeating cycle seldom occur is that the
forces between atoms that are not electrically charged
fall off quite rapidly with distance. If an orderly stacking
with a repeating cycle of five atomic layers, for example, is to exist, the atoms must influence one another

when they are five layers apart. Neutral atoms do not exert much force on one another at such distances.

Indeed, it is surprising how well the orderliness of stacking is maintained in crystals that adopt one of the two simplest close-packed structures. The distinction between the two is not one that affects the relations between adjacent layers, but only the relation between layers separated by an intervening layer. In other words, the distinction cannot be caused by forces between nearest-neighboring atoms, but only between next-nearest neighbors.

In a few crystals the atoms do find it difficult to control the orderliness of stacking. Their arrangement may look as if the stacker forgot his stacking order momentarily. In the middle of an orderly A B A B sequence there may appear a layer of atoms in C-positions; or the arrangement may switch from A B A B to C B C B. Such "stacking faults" are still another possible form of defects in crystals, to add to those discussed in Chapter I. Crystals of metallic cobalt are especially subject to these faults. In fact, cobalt seems to have unusual difficulty deciding which of the two simple close-packed arrangements it prefers.

Close-packed Unit Cells

In picking out suitable unit cells for these two simplest close-packed structures, it is natural to start with a two-dimensional building block for one of the layers, and then to add a thickness dimension in the direction perpendicular to the layers. This chapter has already discussed suitable two-dimensional building blocks for the close-packed structure of atoms in a plane. The hexagonal block (Figure 77) is a good one to begin with.

Since the stacking order repeats itself after two layers have been laid down in the A B structure, and after three layers have been laid down in the A B C structure, you can guess that the unit cell will have a thickness of two layers in the first structure, and of three layers in the second. Such a cell for the A B structure would look like Figure 86.

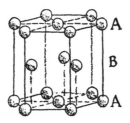

Fig. 86. THE CLOSE-PACKED ARRANGEMENT *at the left of Plate 33 can be built of this hexagonal unit cell.*

Notice that you can choose a smaller cell by starting with a two-dimensional block containing only one atom instead of three, as Figure 80 pointed out. If you pick the two-dimensional block as a rhombus instead of a hexagon, you obtain the unit cell shown in Figure 87.

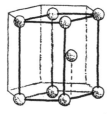

Fig. 87. A TWO-ATOM UNIT CELL *for the first close-packed arrangement in Plate 33 can be chosen from the hexagonal cell of Figure 86.*

It is a remarkably unsymmetrical cell, to be sure, but it is a true unit cell. It assembles the structure correctly when it is repeated so as to fill space, and it pleases the crystallographer, as the hexagonal cylinder does not, be-

cause it is a parallelepiped, a six-sided polyhedron whose
faces are parallel in pairs. The content of the cell,
counted by the method we have described, turns out to
be two atoms. Crystallographers call the structure "hexagonal close-packed." In crystals of the metal magnesium the atoms adopt this arrangement.

Proceeding in the same way with the more complicated looking A B C structure, you extend the cell to a
depth of three layers instead of two and obtain the cells
shown in Figure 88. The parallelepiped in Figure 88B
contains three atoms this time.

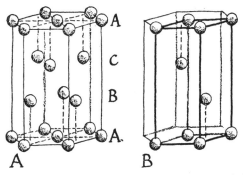

Fig. 88. THE CLOSE-PACKED ARRANGEMENT *at the
right of Plate 33 can be built of these unit cells.*

Examining the symmetry of the cells whose cross-
sections are hexagons, you would be tempted to assign
both structures to the hexagonal crystal system. For the
A B structure that assignment would be correct. But the
A B C structure will surprise you; it has more symmetry
than you would guess at first glance. It is cubic!

The most convenient way to find a cubic unit cell for
the A B C structure is to extend the B-layers and C-
layers in the diagram by including three more atoms
in each, as in Figure 89A. Then you can choose a unit
cell in which all the atoms lie either on the corners or on

CRYSTALS AND CRYSTAL GROWING

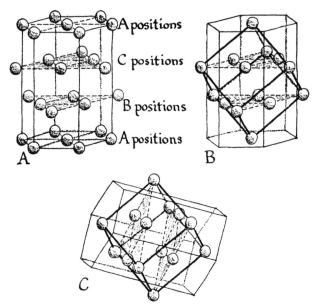

Fig. 89. A CUBIC UNIT CELL *for the close-packed arrangement in Figure 88 is found by adding to the diagram a few more of the atoms in B and C positions (A). Since the structure is cubic (B), the stacking of the spheres (A) could be carried out equally well in any of three other directions, as is shown (C) for one of those directions.*

the centers of the faces of a cube, Figure 89B. For that reason the structure is called "face-centered cubic." You can picture it as in Figure 90A. It is the arrangement adopted by the atoms in a crystal of copper, for example.

To reassure yourself that the face-centered cubic structure is indeed close-packed, it is helpful to choose a unit cell of the kind shown in Figure 90C with one of its atoms at the center of the cell. Then the cell exhibits the atom's twelve neighbors, all at the same distance from it, at the centers of the twelve edges of the cube.

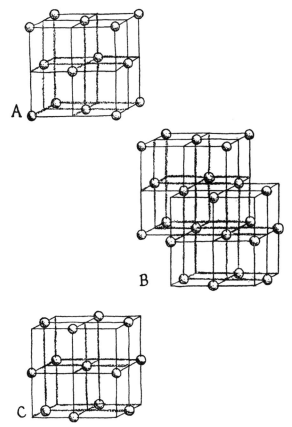

Fig. 90. THE CUBIC CLOSE-PACKED UNIT CELL (*A*)
*of Figure 89 shows why its structure is often called
"face-centered cubic." When the diagram is extended
to include more atoms (B), it is clear that a different
unit cell can be chosen with an atom at the center and
atoms at the centers of the twelve edges. This cell (C)
shows that each atom has twelve immediate neighbors,
all at the same distance.*

In other words, the cell makes it clear that the face-centered cubic structure has the essential property of any close-packed structure: each atom is surrounded by twelve others, all at the same distance.

You will notice an unexpected result of the discovery that the A B C structure is cubic, when you remember that all cubic structures have four equivalent axes of three-fold symmetry. One of these axes is perpendicular to the layers of atoms which you laid down in A B C order in the course of building the structure. But since all four of the three-fold axes are equivalent, you have no way of telling from the finished structure that you stacked the layers perpendicular to a particular one of those axes. You could have built the structure equally well by stacking the layers perpendicular to *any* of the four axes. In Figure 89C you see how the same unit cell as that in Figure 89B, in the same orientation, would serve if you had stacked the layers perpendicular to another of the axes. The same atoms take the same positions, whichever way you choose to stack them; the only difference is in how you think of the "layers."

Even though the face-centered cubic cell contains four atoms—one more than the cell of Figure 88B—it is a more appropriate unit cell because it is the most symmetrical kind that you can pick for the structure. But in studying the physics of crystals, it is often useful to know the minimum size of unit cell for a given structure, even though it may not be the most symmetrical.

There is a way of looking at the face-centered cubic cell in Figure 91A which makes it easy to find a cell of half the size. The atoms in any plane including a face of the cube have the arrangement shown in Figure 91B. In it the smallest two-dimensional building block is the square whose edges are at 45° to the edges of the cubic unit cell. Following this clue, you can extend the three-dimensional structure in the way shown in Figure 91C,

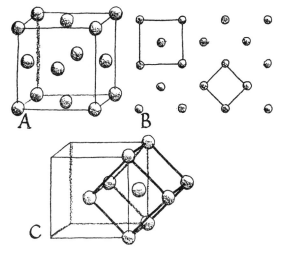

Fig. 91. A TWO-ATOM TETRAGONAL UNIT CELL *can be constructed for the face-centered cubic structure (A), by choosing the smaller two-dimensional building block (B) for the atoms in one cubic face, and extending the block in the third dimension (C).*

and then outline a tetragonal unit cell—a cell having the shape of the building block in the tetragonal crystal system (Plate 28).

Thus, for the face-centered cubic structure, you have found a cubic unit cell of four atoms (Figure 91A), a hexagonal cell of three atoms (Figure 88B), and a tetragonal cell of two atoms (Figure 91C). But even the last is not the smallest cell for the structure. There is a "rhombohedral" unit cell containing only one atom—a cell whose shape is like that of the rhombohedron which you obtained by squashing a cube from corner to opposite corner (Figure 70), except that in this case the cube is stretched, not squashed.

You can outline a rhombohedral cell by starting with the usual face-centered cubic cell of Figure 90A, and drawing lines from an atom at one corner to the atoms at the centers of the three faces that meet at the corner. Then draw similar lines from the atom at the opposite corner to its three similar neighbors. The eight atoms now fall at the eight corners of a rhombohedron, whose outlines can be completed by drawing six more lines, as Figure 92 shows.

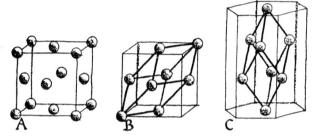

Fig. 92. THE RHOMBOHEDRAL UNIT CELL (*B*) *of the face-centered structure* (*A*) *contains one atom. At C you see the same cell from the "stacking" point of view.*

Other Structures

Turning now to other arrangements atoms adopt in crystals, you will find that your study of close-packed structures has given you helpful ways of looking at structures that are not close-packed. For instance, the rhom-

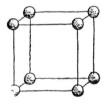

Fig. 93. THE SIMPLE CUBIC UNIT CELL *is a cube containing only one atom.*

bohedral unit cell for copper in Figure 92, which you
have just derived, suggests another structure whose unit
cell would have atoms on the eight corners of a cube
instead of a rhombohedron. That structure, called "simple cubic," is shown in Figure 93. It is certainly remarkable in one respect: no known substance crystallizes that
way.

On the other hand, many metals—iron, for example—
crystallize in a structure whose unit cell is a cube in
which atoms occupy not only the corners but also the
center. Figure 94 pictures this "body-centered cubic"

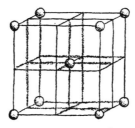

Fig. 94. THE UNIT CELL OF IRON *shows why its structure is called "body-centered cubic." Here each atom
has eight immediate neighbors, all at the same distance.*

structure. It is sometimes convenient to think of the
structure as consisting of two of the simple cubic arrangements pictured in Figure 93. The two "interpene-
trate" each other; each atom belonging to one of them
is surrounded by eight atoms belonging to the other.

Since the unit cell shown in Figure 93 for the simple
cubic structure contains one atom, the unit cell for the
body-centered cubic structure in Figure 94 contains two
atoms. But again you can find a unit cell with only one
atom. As in the case of the face-centered cubic structure, the one-atom unit cell for the body-centered cubic
structure is a rhombohedron, shown in Figure 95B.

Notice that when the atoms adopt the body-centered

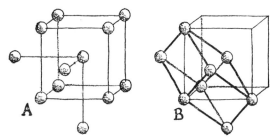

Fig. 95. THE RHOMBOHEDRAL UNIT CELL (*B*) *of the body-centered structure (A) contains one atom.*

cubic arrangement, they are no longer close-packed. Each atom has only eight near neighbors instead of twelve. You can guess that in those crystals the atoms are not simply attracting one another indiscriminately from all directions. There must be other forces tending to make the atoms stand away from one another at definite angles, and thus restrict the number of neighboring atoms which each atom will accommodate.

Already you have seen in Figure 79 an imaginary two-dimensional structure with the same property; each atom is directly associated with relatively few neighbors. A real structure in which you can see this property carried to an extreme is that adopted by the silicon atoms in a silicon crystal, and by the carbon atoms in a diamond. In those crystals each atom has only four near neighbors —only a third of the number that it would have if the atoms were behaving like little spheres which attracted one another indiscriminately.

Figure 96 shows a cubic unit cell for the diamond structure. You can think of the four neighbors of each atom as located at the corners of a tetrahedron surrounding that atom. By dividing the unit cube into eight little sub-cubes (Figure 96A), you can see that the structure is remarkably "open." There are atoms at the centers of four of the eight sub-cubes, and the centers of the other

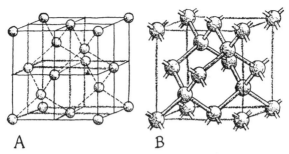

A   B

Fig. 96. THE UNIT CELL OF SILICON *and diamond
(A) shows that the structure is quite "open." Only half
the eight little sub-cubes have atoms at their centers,
and each atom has only four immediate neighbors. In a
diagram (B) which suggests the reason for this open-
ness, each atom is connected to each of its immediate
neighbors by a "stick." In fact each atom is bonded to
four others by bonds at definite angles to one another,
and models of situations like this are often made by
using sticks to represent the bonds.*

four sub-cubes are unoccupied, even though all eight
sub-cubes offer the same amount of space.

Figure 96B suggests why the atoms of this structure
do not pack together closely. The "sticks" connecting
the neighboring atoms point out that those neighbors
are linked by strong bonds at definite angles. The limited
number of bonds formed by each atom, and the definite
angles at which the bonds stand, together make the structure much more "open" than the close-packed structure
of copper.

When you pass on to the arrangements of crystals
with several kinds of atoms, you find that you can profit
from your examinations of structures containing only
one kind. The crystal structure of sodium chloride
(Figure 13) is a good example. Here each ion has six
nearest neighbors of the opposite sign, and twelve next-
nearest neighbors of the same sign. The ions of either

sign separately form a face-centered cubic arrangement.
But, of course, that arrangement is not literally close-
packed. The ions of the same sign are repelling one
another, not attracting one another as neutral atoms do
in the close-packed structures. Indeed, as Chapter I described, those ions are held together only because their
arrangement is interpenetrated by a similar arrangement
of ions of the opposite sign.

The fact that you can consider the atomic arrangement in sodium chloride as two interpenetrating face-
centered cubic structures gives you an easy way to find a
unit cell with only one sodium ion and one chloride ion
—a unit cell of the smallest possible size. You have already found such for the face-centered cubic structure,
in Figure 92. You can draw that cell for the sodium ions
alone, and then put the chloride ions in their proper
places. It turns out that just one chloride ion falls inside
the cell. Figure 97 shows the resulting rhombohedral unit

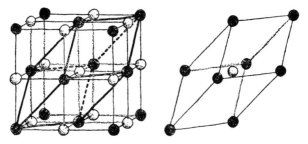

Fig. 97. THE RHOMBOHEDRAL UNIT CELL *of the sodium chloride structure contains one atom of each sort.*

cell for sodium chloride, with chloride ions at the corners
and a sodium ion at the center. Of course, you could
just as well choose a rhombohedral cell with sodium ions
at the corners and a chloride ion at the center.

As you look at these diagrams of structures, you must
recall every once in a while how abstract they are. The

little circles represent the positions of atoms which "touch" one another. Of course, atoms are not really hard spheres, but in crystals they often act somewhat as if they were. When a crystal contains atoms of different species, the different atoms may behave as if their sizes differed also. In sodium chloride, the spheres representing the chloride ions are larger than those representing the sodium ions, as the "packing model" of the sodium chloride structure in Plate 34 suggests.

The structure of cesium chloride crystals, shown in Figure 98, is another adopted by ionic crystals when the

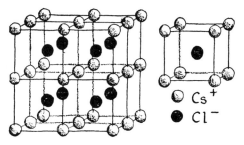

Fig. 98. IN THE CESIUM CHLORIDE STRUCTURE *each ion is surrounded by eight oppositely charged, at the corners of a cube.*

ions of the two kinds are present in equal numbers. Here each ion has eight nearest neighbors of the opposite sign, and six next-nearest neighbors of the same sign. The cesium chloride structure can be thought of as like the body-centered cubic structure shown in Figure 94 and Plate 35, but with the two different sorts of ions alternately occupying the sites in the structure. Or you can think of the structure as two interpenetrating simple cubic structures, in the same way that you can think of the sodium chloride structure as two interpenetrating face-centered cubic structures.

Another structure you can consider as two interpenetrating face-centered cubic structures is that of zinc sulfide in the form of the mineral sphalerite. As Figure 99 shows, the zinc atoms and the sulfur atoms alternately occupy atomic sites arranged like those of the atoms in the diamond structure (Figure 96).

Even though this arrangement is like that of sodium chloride in consisting of two face-centered cubic struc-

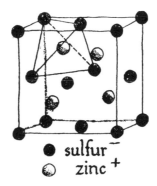

● sulfur⁻

◎ zinc⁺

Fig. 99. IN THE SPHALERITE STRUCTURE *each zinc atom is tetrahedrally surrounded by four sulfur atoms.*

tures, there is an important difference from the point of view of the ions. In the sphalerite structure each atom has only four nearest neighbors. If the atoms were ions of two oppositely charged kinds attracting or repelling one another indiscriminately, they would surely rearrange themselves so that each acquired more oppositely charged neighbors, as in the sodium chloride and cesium chloride structures. You can infer that the zinc and sulfur atoms in sphalerite are not acting as if they were simply ions. To be sure, the zinc and sulfur atoms acquire positive and negative charges; but they decide to stand away from one another at definite angles because they also form chemical bonds, as the carbon atoms in diamond do. In a later chapter you will see how the resulting structure of sphalerite makes possible a curious electrical effect in that mineral.

There is another arrangement closely related to the diamond and sphalerite structures, however, in which the atoms behave more nearly like simple ions. That is the structure of the mineral fluorite. Fluorite is calcium fluoride, made of positively charged calcium ions associated with twice as many negatively charged fluoride ions. Since the negative ions have only half as many positive ions with which to content themselves, it is not surprising that each fluoride ion has only four calcium ions as nearest neighbors; whereas each calcium ion has eight fluoride ions as nearest neighbors.

Figure 100 shows the fluorite structure. It is a face-

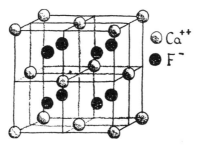

Fig. 100. THE FLUORITE STRUCTURE *embodies twice as many fluoride as calcium ions. Each calcium ion has eight fluoride neighbors at the corners of a cube, and each fluoride ion has four calcium neighbors at the corners of a regular tetrahedron.*

centered cubic structure of calcium ions, interpenetrated by a simple cubic structure of fluoride ions. The eight fluoride ions immediately surrounding each calcium ion are at the corners of a cube with the calcium ion at the center. The four calcium ions surrounding each fluoride ion are at the corners of a regular tetrahedron with the fluoride ion at the center. You can think of the fluorite structure as like the sphalerite structure (Figure 99), but with all the little sub-cubes (Figure 96) filled with fluoride ions.

When a substance is made of many different sorts of atoms, the structure of its crystals will necessarily be more complicated than those you have been examining. You cannot avoid this conclusion when you remember that the smallest unit cell that you have found for any material has always contained at least as many atoms as the chemical formula for the substance. The chemist would write copper "Cu," and you can find a unit cell for a copper crystal with only one atom. The chemical formulae NaCl for sodium chloride and CaF_2 for fluorite tell you at once that unit cells for their crystals must contain at least two atoms and three atoms, respectively.

In other words, the unit cell must contain at least one "formula unit" of the crystal substance. If it contains even more, then it must contain some multiple of a formula unit. Looking back at the chemical formulae for the crystals you have grown, you can see that the unit cells for most of them must contain a great many atoms. The formula unit of alum, for example, has forty-eight atoms, and the smallest unit cell contains four of these formula units. It is remarkable, indeed, that progress in the interpretation of X-ray diffraction pictures has reached a stage where it is possible to find such complicated atomic patterns in three dimensions.

Some complicated structures can be pictured in terms of simpler ones, much as the study of close packing has led you in this chapter, by way of the face-centered cubic structure and the diamond structure, to understand the structure of sphalerite. In the next chapter you will see how the structure of sodium nitrate and calcite crystals can be pictured in terms of the structure of sodium chloride, and how that picture enables you to understand some of the physical properties of those crystals.

PROBLEM 7

1. Show that stretching the simple cubic structure in Fig. 93 by the right amount along any of its axes of three-fold symmetry, converts it into the face-centered cubic structure.

2. Show that not only the face-centered cubic structure but also the body-centered cubic structure can be converted to the simple cubic structure, by squashing it or stretching it to the right extent.

3. Show that the face-centered cubic structure can be converted to the body-centered cubic structure by squashing it either along an axis of three-fold symmetry or along an axis of four-fold symmetry.

CHAPTER IX

Cleaving and Gliding Crystals

A kind of behavior found only in crystals is "perfect cleavage": a crystal may break apart much more easily along planes in a few directions than along any others. Not all crystals exhibit cleavage; but in those that do, the property is an impressive testimonial to the orderliness of their atomic arrangement. A familiar example of perfect cleavage among minerals is the cleavage of mica, shown in Plate 3. Among the crystals you can grow by the recipes in this book, sodium nitrate, nickel sulfate hexahydrate, and calcium copper acetate hexahydrate cleave beautifully, and copper acetate monohydrate cleaves fairly well.

The plane of cleavage in a crystal of nickel sulfate hexahydrate is perpendicular to its axis of four-fold symmetry. You can cleave the crystal along any plane perpendicular to that axis: the term "cleavage plane" does not refer to a particular place in the crystal but to a particular direction in the crystal. The crystal can be cleaved along any one of the entire family of planes in that direction. As with mica, you can cleave plates whose thinness is limited only by your skill (see Plate 36).

In sodium nitrate, the cleavage is rhombohedral. The natural faces of a sodium nitrate crystal form a rhombohedron. Parallel to each natural face there is a family of cleavage planes, and thus there are three such families in sodium nitrate (Plate 36) instead of the one family characteristic of nickel sulfate hexahydrate.

One consequence of the rhombohedral cleavage of sodium nitrate is that the crystal can be cleaved into little rhombohedral blocks. The corresponding property in calcite led scientists of about three centuries ago to think that calcite might be constructed of identical tiny blocks, having the shape of a cleavage rhombohedron. This guess made the beginning of the building-block theory of crystals outlined in Chapter VI. The last chapter has described the fruitfulness of this theory, and its reconciliation with today's picture of a crystal as built of a repetitive orderly array of atoms.

Cleaving a crystal cleanly is an art and takes practice —one of the least of the arts the gem cutter must learn. You must resign yourself to destroying a few crystals before you master it. Put the crystal you want to cleave on a table, holding it in a suitable position with a little smear of Plasticine. Place the sharp edge of a single-edged razor blade along the plane of cleavage. While you hold the blade in position, give it a light tap with something like a kitchen knife or a screwdriver—something heavier than a pencil and lighter than a carpenter's hammer. The crystal will come apart along the cleavage plane and present a fresh flat face.

After you have mastered this art, you can easily assure yourself that the cleavage directions are very specific. An attempt to cleave the crystal in some other direction will produce only irregularly broken surfaces. Looking carefully at these broken surfaces, however, you may find small, brightly reflecting sections which lie along planes of cleavage. This is one way to locate the cleavage directions, if there are any, in a crystal you have never met before.

Notice that cleavage has nothing to do with imperfections in the crystal. Indeed, if you try cleaving an imperfect crystal, you may find that it cannot be cleaved as well as a perfect one. The cleavage is a result of the orderliness of the atomic arrangement. No wonder imperfections, which disturb the orderliness, disturb the cleavage too.

You will note also that the cleavage directions are obedient to the symmetry of the crystal. Of course, symmetry alone cannot dictate in what direction a crystal will cleave, nor indeed whether it will cleave at all. But symmetry does demand that, if the crystal cleaves in some direction, then it must also cleave in just the same way in any other direction symmetrical to the first direction.

The cleavage of sodium nitrate illustrates this very well. The rhombohedral sodium nitrate crystal has a three-fold axis of symmetry along the direction of the short diagonal of the rhombohedron, as shown in Figure 101. Consequently, the family of cleavage planes parallel to any face must be duplicated when the crystal is turned one third of a revolution about that axis in either direction. The three families of indistinguishable cleavage planes fulfill this requirement.

It is interesting to contrast the cleavage properties of calcium copper acetate hexahydrate with those of nickel sulfate hexahydrate. Each of these crystals belongs to the tetragonal crystal system, and each has directions of perfect cleavage. In nickel sulfate the family of cleavage planes is *perpendicular* to the four-fold axis of symmetry, and the symmetry does not require that family to be duplicated in any other direction. In calcium copper acetate, however, the cleavages are in planes *parallel* to the four-fold axis. Hence, each family of cleavages is dupli-

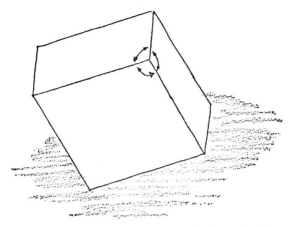

Fig. 101. THE RHOMBOHEDRON, *which is the shape of a sodium nitrate crystal, is a cube that has been squashed down along the direction of one of its three-fold axes. There are two corners at which all angles are obtuse (greater than 90°), and the axis lies along the line between those corners. The cubic sodium chloride crystal can be cleaved in the planes parallel to the cube faces. Similarly, and for much the same reasons, sodium nitrate can be cleaved in planes parallel to the rhombo-hedron faces.*

cated by another family at 90° to it—one quarter of a turn about the axis. In fact, you can cleave thin single-crystalline needles, directed along the axis, out of a single crystal of calcium copper acetate. The needles do not break easily because there is no cleavage plane cutting across the axis.

When a crystal has two directions of cleavage not related by symmetry, both may be duplicated by symmetry in other directions. Again calcium copper acetate is a good example. One of the families of cleavage planes parallel to the axis affords very good cleavage. These planes are parallel to the larger faces in the usual habit of the crystal. There is another family, also parallel to the axis but at 45° to the first, in which cleavage is poorer. The cleavage surfaces show threadlike lines parallel to the axis. Since each of these families is duplicated by symmetry at 90°, you can turn the crystal about the axis in steps of 45°, obtaining alternately good and poor cleavages.

Without doubt, the planes of cleavage in a crystal are planes across which the forces between the atoms—the forces binding the crystal together—are weakest. It is tempting to guess that, if the binding forces are weakest in the direction perpendicular to those planes, then the atoms are farther apart in that direction than in others. But common salt—sodium chloride—shows that the atomic behavior cannot be as simple as that. Salt has cubic cleavage—three mutually perpendicular families of

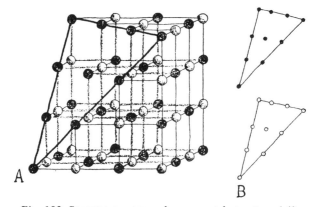

Fig. 102. SODIUM CHLORIDE *has a crystal structure (A) in which any plane of atoms parallel to a cube face contains equal numbers of both sodium and chloride ions, but any octahedron plane contains only one kind of ions. Two adjacent octahedron planes (B) have op- posite electric charges; two adjacent cube planes are both neutral.*

cleavage planes. To assert that the atoms were farther apart *perpendicular* to a cleavage plane in one family would imply that they were also farther apart *within* a cleavage plane in another family.

As a matter of fact, sodium chloride provides a good example for study of the atomic picture of cleavage. Its arrangement of positively charged sodium ions and negatively charged chloride ions, already shown in Figure 13, is easy to understand. First pick out of this structure, shown again in Figure 102, one of the "cube" planes of ions—one of a family of cleavage planes. Then pick out for comparison one of the "octahedron" planes of ions—a plane along which the crystal does not cleave.

The arrangement of ions in the cube plane is like a square net with positive and negative ions located alternately in its meshes. The ions in the octahedron plane are arranged on a net with triangular meshes, and all the ions are either positive or negative. Since there is the same number of positive and negative ions in the cube plane, the electric charge in the whole plane adds up to zero. The plane is electrically neutral over-all, and so are the parallel planes on either side of it.

But an octahedron plane, whose ions are all one sort, has a large electric charge, and the parallel planes on either side of it have a large charge of the opposite sign. As you know from the discussion of electric forces in Chapter I, the planes of opposite charge will attract each other much more strongly than the neutral planes. Therefore, the crystal comes apart much more easily between the neutral planes than between the charged planes.

This examination of the reason why sodium chloride shows cubic cleavage makes clear also why sodium nitrate shows rhombohedral cleavage. Studies of the structure of sodium nitrate by X rays reveal that its atomic arrangement is somewhat similar to that of sodium chloride, as Figure 103 shows. In place of the spherical

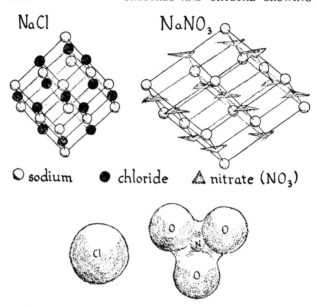

Fig. 103. SODIUM CHLORIDE AND SODIUM NITRATE
*have similar crystal structures. But the nitrate ion is
not spherical, like the chloride ion. Its roughly triangular shape accounts for the fact that the structure of sodium nitrate is a distortion of the structure of sodium
chloride. The rhombohedral cleavage of sodium nitrate
(and of calcite) therefore has the same atomic explanation as the cubic cleavage of sodium chloride.*

chloride ions, sodium nitrate contains nitrate ions. The
four atoms constructing a nitrate ion are all in the same
plane, and each ion can be roughly pictured as a triangular block. The blocks are arranged in parallel planes,
and the cubic structure of sodium chloride is broadened,
in the plane of the blocks, into a rhombohedral structure. As in sodium chloride, the electric forces ex-
erted by the positively charged sodium ions and the negatively charged nitrate ions have caused them to collect

their opposites around them and to push their fellows away. The rhombohedral planes are electrically neutral in sodium nitrate, just as the cubic planes are neutral in sodium chloride.

Another crystalline phenomenon, much less frequently met than cleavage, is "glide." The mineral calcite is the most celebrated exhibitor of glide; and since sodium nitrate crystals have the same atomic arrangement as calcite, you can observe glide in the crystals you have grown.

Glide occurs when you press the edge of a knife into one of the edges of a rhombohedron of sodium nitrate. You must choose an edge where the faces meet in an obtuse angle rather than an acute angle. In order to obtain the best result, choose a place on the edge not more than an eighth of an inch from one end—the end where the edge makes an obtuse angle with the face cutting it off. As you press the blade slowly into the crystal, the

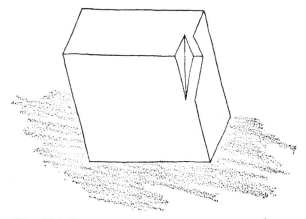

Fig. 104. GLIDE IN SODIUM NITRATE *occurs when a knife is pressed into a suitably chosen edge of the crystal. Find such an edge by comparing your crystal with this picture and with Figure 101.*

part of the crystal between the blade and the end of the edge will shift over as in Figure 104.

Like cleaving, gliding takes practice. Press gently at first; as glide progresses, press increasingly hard because you are shifting the material of the crystal along a progressively larger surface. Keep the edge of the blade parallel to the surface along which the crystal is gliding —in other words, at equal angles with the two faces at the edge into which you are pressing the blade. After a successful performance, the part that has glided still looks fairly clear, and the new face at the end is quite flat.

Glide, as you can see, is entirely different from cleavage. Sodium nitrate cleaves along certain planes and glides along entirely different planes (Figure 105)—

Fig. 105. THE THREE GLIDE PLANES *in sodium nitrate. The families of glide planes, which are parallel to these planes, are not in the same directions as the three families of cleavage planes, shown in Plate 36.*

planes just as well defined as the cleavage planes, but not the same. From the appearance of the glided part you can guess that its atomic arrangement must be like that of the original crystal. The glided piece looks like an image of part of the main crystal.

From these observations you can have some confidence in making a picture of what happens on an atomic scale. There are planes along which the atoms can slide fairly freely. They do not need to slide far before they find that they are in positions that look much like their old positions, and they are content to settle down in the new ones.

Plate 23. Crystal of Strontium Formate
Dihydrate (colorless).

Plate 24. Crystal of Lithium Trisodium
Chromate Hexahydrate (yellow).

Plate 25. Crystal of Calcium Copper
Acetate Hexahydrate (blue).

Plate 26. "Hopper growth" is a spectacular kind of facial development, occurring when the crystal grows more rapidly at the edges than at the centers of the faces. It sometimes appears on bismuth crystals grown rapidly from the molten metal (top), and on sodium nitrate crystals grown rapidly from solution (bottom).

Plate 27. Different habits of natural quartz may have been caused by different impurities in the solution from which they grew. Long crystals like that at the left grow from the walls of fissures in the rock at Hot Springs, Arkansas. Short "doubly terminated" crystals are characteristic of the "Little Falls diamonds": quartz found in Herkimer County, New York.

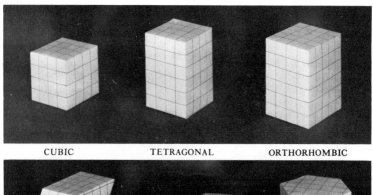

CUBIC   TETRAGONAL   ORTHORHOMBIC

MONOCLINIC   TRICLINIC   HEXAGONAL

Plate 28. The six crystal systems. There are six styles of crystalline architecture, each employing a different sort of building block; and every crystal belongs to one of these six "crystal systems." Lines are drawn in these models to assist in visualizing which angles are right angles and which are not. In the CUBIC block all angles are right angles and all sides are equal. In the TETRAGONAL block all angles are right angles but there are two different lengths of side. In the ORTHORHOMBIC block all angles are right angles but there are three different lengths of side. The MONOCLINIC block is like the orthorhombic block, pushed so eight of its angles are no longer right angles. In the TRICLINIC there is *no* right angle and there are three different lengths of side. The HEXAGONAL block is a hexagonal prism, with right angles between its vertical sides and its top and bottom faces.

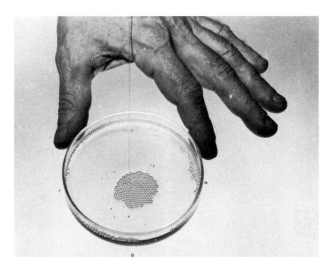

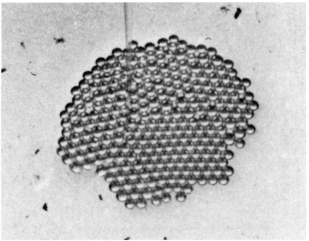

Plate 29. Soap bubbles form in rafts on the surface of a soap solution. Their arrangement is close-packed in two dimensions, and shows "grain boundaries" (Fig. 10).

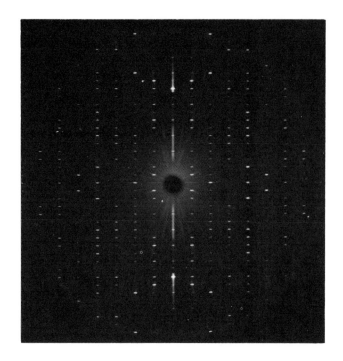

Plate 30. Typical X-ray photograph of a crystal. A single beam of X rays is scattered into many beams, which make the spots on the photograph. The unscattered part of the beam made the large spot in the middle.

Plate 31. A model of an arrangement of atoms in three dimensions obscures the arrangement because the "atoms" hide one another.

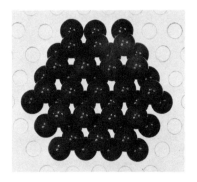

Plate 32. Two close-packed layers of marbles. Each sphere in the second layer touches six spheres in its own layer and three in the first layer, and sits over an open space in the first layer. [In order to show the structures more clearly, the marbles in these photographs which are described as "touching" have been spaced out uniformly.]

Plate 33. Alternative arrangements of three close-packed layers of spheres. At left, the spheres in the third layer are directly over the spheres in the first layer. At right, the spheres in the third layer are over open spaces in the first layer. In both arrangements each sphere in the second layer touches a total of twelve other spheres.

Plate 34. "Packing model" of sodium chloride. As the model of how the ions pack together suggests, the negative ions are larger than the positive ions.

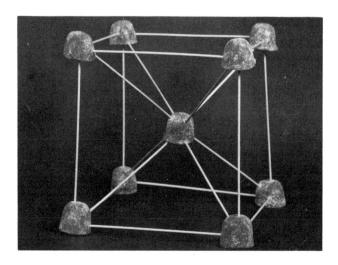

Plate 35. The body-centered cubic structure, already shown in Fig. 94. Gumdrops and toothpicks furnish convenient means for making models of crystal structures. You will find that the construction of such models helps you in understanding the structures that are difficult to visualize, such as the diamond structure (Fig. 96).

A

Plate 36. Cleavage of sodium nitrate occurs in the three families of planes parallel to the natural faces of the rhombohedral crystals. It is often a help in removing cloudy parts before using the crystals in optical experiments. Part A

B

C

D

shows two cleavage plates taken from the crystal, with two of the three families used. Cleavage of nickel sulfate hexahydrate occurs in the family of planes perpendicular to the axis of four-fold symmetry (part B). Calcium copper acetate hexahydrate cleaves well in two families of planes (part C) and less well in two others (part D).

Plate 37. A twin of ammonium dihydrogen phosphate, which grew accidentally in the laboratory.

Plate 38. Apparatus for demonstrating the piezoelectric effect (Fig. 112). The lamp flashes when you rap the crystal with a hammer.

Plate 39. A pencil line viewed through calcite appears to be doubled.

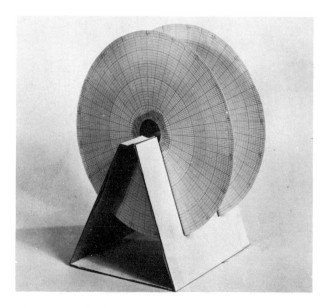

Plate 40. A polarimeter you can easily make (Fig. 124).

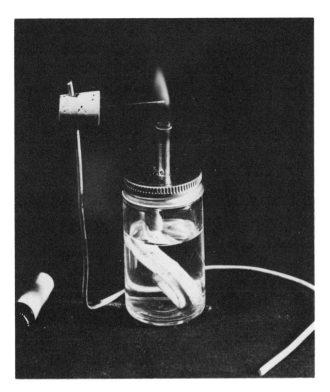

Plate 41. Alcohol lamp.

Plate 42. Wooden spectroscope.

By pushing on the knife blade, you urge first one plane of atoms and then the next to take this small trip. You are never urging a volume of atoms to move at once; at any one instant you are urging only a plane of atoms to move. The moving plane carries on its back all the planes that have already taken the plunge (Figure 106).

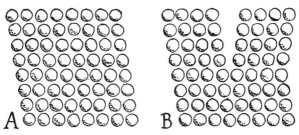

Fig. 106. IN THE GLIDED STRUCTURE (B) of a crystal, successive planes of the atoms in the unglided structure (A) have moved a short distance, and the atoms have settled in new positions symmetrical to the old positions.

Crystals of ammonium nitrate glide spectacularly. This substance forms long needle-shaped crystals. Operating on one of these needles slowly, you can bend it into a circle, twist it into a spiral, or even tie a knot in it.

Since metals are also made of crystals, and since they can often be deformed mechanically as easily as ammonium nitrate, it is natural to visualize the bending of a metal as a gliding process in its constituent crystals. Single crystals of metals do in fact have well-defined glide planes, and glide is responsible for many of their mechanical properties. The study of metals has shown that it is not necessary for a whole plane of atoms to move at any one instant in order to accomplish glide. It is only necessary to move a line of atoms, then the next line in the plane, and so on. In fact, the study has shown that glide would almost certainly not occur if an

entire plane of atoms had to move at once, because
it would be too difficult to move so many atoms si-
multaneously.

You may ask how it is possible for a line of atoms to
move into another position without moving the next line
of atoms out of its way, and so moving at one time all
the lines of atoms that make up the plane. The answer
has been found in the fact that almost all crystals contain the defects called "dislocations," which Chapter I
mentioned. Dislocations in a crystal permit atoms to
shift their positions line by line instead of plane by plane,
and make glide possible.

The explanation of the cleavages in sodium chloride
and sodium nitrate, in terms of the forces of repulsion
and attraction between the like and the unlike electrical
charges, is a satisfactory one for crystals composed of
ions. Clearly this explanation is hard to use when the
atoms are held together by chemical bonds. But in such
cases cleavages can sometimes be explained in a different way. The cleavage planes can be pictured as those
across which the fewest bonds need to be broken.

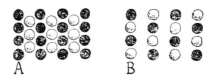

PROBLEM 8

1. In a two-dimensional crystal with one of the above
   structures, you might expect cleavage and glide to occur along the same planes. In a crystal with the other
   structure you might expect cleavage and glide to occur along different planes, as in calcite and sodium
   nitrate. Which is which?

2. When a crystal is not ionic, it is nevertheless sometimes possible to guess, from a knowledge of its
structure, in what direction it might cleave. Diamond,
for example, is held together not by the attractions between oppositely charged ions but by a network of
chemical bonds between its carbon atoms (Figure
96). Decide whether diamond would be more likely
to cleave parallel to its octahedron faces or its
cube faces by finding the ratio of the number of
bonds which must be broken per unit area in the two
cases.

By making a part of the sodium nitrate crystal glide,
you produce an example of another frequent phenomenon in crystals—"twinning." The glided part and the part
directly beneath it, the part that has not yet glided but
would if you pushed the blade farther, taken together

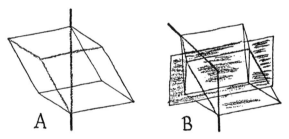

Fig. 107. A TWINNED CRYSTAL (*B*) *of sodium nitrate results from glide. The parts of the twin are symmetrically related by a plane of symmetry at the twin
boundary. Both parts of the twin have the same properties and the same symmetry, but the symmetry operations of the two parts no longer coincide. For example, the axis of three-fold symmetry of the untwinned
crystal (A) is aligned in different directions in the two
parts of the twinned crystal.*

constitute a twinned crystal. There is a "twin boundary" along the plane of atoms directly beneath the last plane that slipped, and the two parts of the twin are mirror images of each other in that plane. Figure 107 shows the plane of reflection for such a sodium nitrate twin.

In this case you have produced a twin by glide, but twinning in general is a much more prevalent phenomenon than glide. The word is used whenever two or more parts of a crystal are tightly joined and have the same atomic arrangement, but differ in orientation in such a way that they are symmetrically related to one another. The two parts of the twinned sodium nitrate crystal are symmetrically related by reflection across the twin boundary. A similar plane of reflection relates the two parts of the twinned crystal of ammonium dihydrogen phosphate shown in Plate 37, which grew spontaneously in a tank in which many untwinned crystals of the same substance grew at the same time.

CHAPTER X

Melting and Transforming Crystals

One of the touchstones to the true solid—a crystalline solid, in contrast to a merely "rigid" glass—is its sharp melting temperature. When you heat the crystal slowly, it remains rigid until it reaches its melting point. Then, if you keep it at that temperature, it will melt completely to a liquid. The temperature need not increase as the crystal melts; the melting point is a critical temperature for it, and there its orderliness suddenly collapses into disorder.

As you heat a glass, on the other hand, it becomes softer and softer; there is no critical temperature at which it suddenly melts. It already has a disordered structure, and the rising temperature simply permits its molecules to move past one another more and more readily. There are many little orderly patches in both a glass and a liquid, but their orderliness extends over only the short range of a few hundred atoms, in contrast to the "long-range order" of a crystal, which extends over millions of atoms.

Of course, there are crystalline solids whose crystallinity cannot be verified by observing sharp melting points, because the solids decompose chemically when

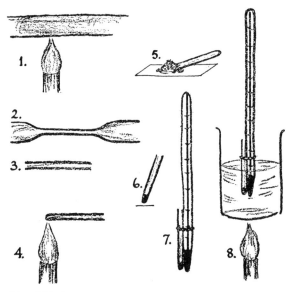

Fig. 108. To determine the melting point *of a crystalline material, prepare a tiny test tube in this way. (1) Heat a piece of quarter-inch Pyrex glass tubing in the flame of a Bunsen burner until it glows dull red and softens. (2) Quickly pull it from both ends to form a thin-walled capillary section of tubing about one-sixteenth inch in diameter and lay it carefully on a noninflammable surface to cool. (3) Break off about an inch of the capillary section, scratching the glass with a triangular file to fix the breaking point. (4) Hold one end of the capillary tube in the Bunsen flame until it softens and seals at the end. (5) Powder some of the crystal, and scoop a little of the powder into the open end of the capillary test tube. (6) Jiggle the powder to the bottom of the tube by tapping it vertically on the table top. (7) Fasten the capillary tube to the side of a thermometer with a rubber band so that the powder is opposite the thermometer bulb. (8) Immerse the thermometer partly in a beaker of liquid, and heat the liquid*

they are heated. And there are a few other exceptions to the rule: rare oddities whose molecular arrangements acquire some disorder at temperatures below the point at which they fall apart into a liquid, and odd rarities that retain some long-range order above the point at which they melt. But in general the test is a reliable one.

To measure the melting point of a solid accurately offers some experimental difficulties. When you heat a solid below its melting point, the heat which the solid absorbs raises its temperature. When the solid reaches its melting point, it continues to absorb heat but it stays at the same temperature until it has melted completely. After it has all liquefied, the liquid will again rise in temperature as it absorbs heat. A great deal of heat is required to melt ice, at 0° centigrade, into water at the same temperature.

You might think that the constancy of temperature during melting would make it especially easy to measure the melting temperature. But the temperature is constant only right next to the melting surface of the solid. Within the solid the temperature is lower than the melting point, because it takes time for the heat to penetrate the solid. And in the surrounding liquid the temperature is higher than the melting point, because the only way to conduct heat constantly to the melting surface is to keep the surrounding liquid hotter than that surface.

A way to surmount this experimental difficulty is to use a very small sample of crystal. Then it will absorb so little heat when it melts that it will all melt nearly at

---

slowly. When you see the opaque powder melt to a clear drop, quickly read the temperature. A preliminary run will locate the melting point approximately. In a second run, with a fresh sample in another capillary tube, you can explore the temperature more slowly and accurately in the range near the temperature you first located.

one time, at a temperature close to that of its environment. If you put the crystal in a "heat bath" made of some liquid that will not react chemically with the crystal and not boil before it reaches the melting point of the crystal, you can slowly heat the liquid, and quickly read its temperature when the little crystalline sample melts.

Chemists often use a simple method of this sort, shown in Figure 108, which will easily give the melting temperature with an error of less than one degree centigrade. Salol is a good crystalline solid on which to try the method. Since its melting point is below the boiling point of water, the heat bath can be made of water.

Crystals of double compounds, such as alum and Rochelle Salt, sometimes behave in a confusing way when they melt. The double compound may begin to melt in the usual way; but if its two ingredients have a higher melting temperature, they will segregate and then solidify again in separate crystals of the two materials. If the double compound is a salt hydrate, you can think of the material as first dissolving in its own water of hydration, and then crystallizing out of that water in a different form.

The double compound, Rochelle Salt, sodium potassium tartrate tetrahydrate, melts at 55° centigrade. Crystals of potassium tartrate hemihydrate and sodium tartrate dihydrate grow in the molten material, and there is a little water left over. The liquid becomes cloudy with the growth of fine needles of the separate tartrates. You can observe this phenomenon by the procedure described in Figure 109. Processes akin to this may often have gone on in the earth, changing the forms taken by the substances composing our planet.

The collapse of order into disorder on melting may not seem unusual when you remember that the atoms in a solid are always vibrating about their orderly positions. When the temperature increases, they vibrate more

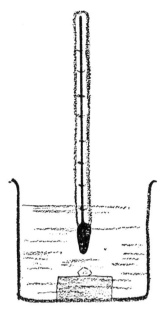

Fig. 109. To MELT ROCHELLE SALT, *put one of the small seed crystals on a metal block in a beaker of mineral oil, which will not dissolve the crystal. Heat the oil slowly, stir it gently with a thermometer, and touch the crystal occasionally with the tip of the thermometer bulb to detect whether it has melted. When it melts, the crystal will tend to stick to the thermometer; you will pull up a thin thread of molten Rochelle Salt as you withdraw the thermometer from the crystal. In this way you can read the melting temperature with an accuracy of about one-half degree. If you then hold the oil a little above this temperature, you will see the crystal melt completely, becoming cloudy as it does.*

vigorously, and finally they are able to move past one another and leave their proper places in the orderly arrangement.

But this picture of melting is not detailed enough to

explain its most characteristic feature—its suddenness. The same picture explains why a glass becomes softer as it becomes hotter. The higher the temperature of a glass, the more readily and the more frequently its atoms move past one another; but there is no critical temperature at which their ability to move changes abruptly.

When a crystal melts, there may be a co-operative action of its atoms. The ability of one atom to move may increase the ability of its neighbors to move. The first few atoms to move are probably vibrating more vigorously than the rest. Perhaps when those few atoms move, they open up just enough space to let some less vigorously vibrating neighbors move. Then the motion of those neighbors would in turn permit even the sluggards to move.

Co-operative action of this sort would explain the difference in the melting of a crystal and the softening of a glass. In a crystal all atoms of the same kind are vibrating about positions of the same kind. In a glass the atoms are vibrating about positions of differing kinds. The atoms vibrating vigorously enough to move in a glass will be those that occupy the kind of positions from which it is easiest to move. Their motion will not greatly assist their neighbors to move, because those neighbors occupy positions from which it was always harder for them to move.

In the crystal, on the other hand, the positions do not favor the motion of one atom over the motion of any other. It is only a little extra vibrational activity in an atom that momentarily favors it. In the picture of co-operative action, the motion of that favored atom then releases the rest, by a process somewhat like a chain reaction.

Many solids show other kinds of behavior, besides melting, that are best explained by visualizing similar atomic co-operation. For example, there are many solids whose atoms can adopt more than one kind of orderliness. In such solids, one orderliness is usually more stable above some definite temperature, another below it. As the solid is heated through that temperature, the crystal structure may change abruptly from one form to the other.

Sometimes the atoms will not shift over to the other arrangement, even when they would find that arrangement more stable. Pure carbon is known in two crystalline forms, graphite and diamond. Graphite is more stable than diamond at all temperatures, at ordinary pressure. But, fortunately for lovers of diamonds, they do not change spontaneously into graphite crystals. In a way not yet understood, some of the carbon in nature crystallized as diamonds, and now those carbon atoms are frozen in that arrangement and cannot shift over into the more stable arrangement (Figure 110).

Quartz is the stable form of silicon dioxide at ordinary temperature. But silicon dioxide adopts many crystalline structures, each most stable within some range of temperatures. As the changing temperature passes through the dividing points between these ranges, some of the transformations occur instantly, others sluggishly, and others apparently never. Whether the transformation does occur depends on whether one structure is closely similar to the other or very different—whether or not the atoms must move far.

Of course, not only heating a solid but also cooling it may bring about a transformation into a more stable form. Many of the pipes in old church organs were made of tin. Several centuries ago a shiny organ pipe in an unheated church would sometimes crumble to gray dust during a winter. "Tin pest" was finally diagnosed as a transformation of tin from its usual crystal structure, belonging to the tetragonal crystal system (Plate 28), to a cubic structure in which tin becomes more stable below

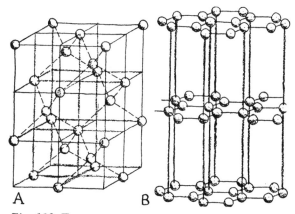

Fig. 110. TWO CRYSTAL STRUCTURES OF CARBON. *The arrangement of atoms in diamond (A) puts it in the cubic crystal system. Each carbon atom is tightly bonded to four others (Figure 96). You can think of any one atom as at the center of a regular tetrahedron, with four other carbon atoms at the four corners. In graphite (B) the carbon atoms are arranged in plane hexagonal nets; the crystal belongs to the hexagonal system. Each atom has only three near neighbors within its net, and the nets are spaced more than twice as far apart as the atoms within a net. Graphite cleaves extremely easily in the plane of the nets; diamond cleaves with much more difficulty but quite smoothly along "octahedral" planes.*

18° centigrade. In fact, it is the same as the structure of diamond, pictured in Figure 110.

You can observe a beautiful example of spontaneous transformation in mercuric iodide. It is especially spectacular because the change of crystal structure brings with it a change in color from red to yellow. The red tetragonal form, stable at room temperature, changes to a yellow form belonging to the orthorhombic crystal system (Plate 28) as the temperature rises through 126°

centigrade. On cooling, the yellow form transforms more slowly back to the red. If you are cautious, you can do this experiment safely by warming a small amount of powdered mercuric iodide in a test tube. The poisonous vapor of mercuric iodide will be largely confined to the test tube, but take care not to breathe it.

Notice that the downward transition in mercuric iodide—from its high-temperature form to its low-temperature form—tends to be slower than the upward transition when it is heated. This fact is reminiscent of the behavior of a solid at its melting point: it usually melts more rapidly than it freezes. You may find that the downward transition must be "seeded," by touching the yellow form with a bit of the red, in much the same way that melted materials can sometimes be supercooled. Tin transforms even more reluctantly when it is cooled, and no doubt this fact preserved many an ancient organ pipe.

Occasionally you may find that when you heat the red form it does not change to the yellow form at the transformation temperature. In other words, you can sometimes superheat a solid above a transformation temperature, but you can never superheat it above its melting point. Undoubtedly this difference in behavior is related to the fact that melting is a change of order into disorder; the structural transformation of a solid is a change of one form of order into another form of order. As Chapter II pointed out, there is seldom any difficulty in attaining disorder but there is often difficulty in attaining order.

When we visualize what the atoms must do to accomplish a structural transformation of a solid, it seems even clearer than in visualizing melting that co-operative action of the atoms might be necessary. In the structural transformation the atoms must not only move; they must move to the new "right places." Since the locations of its neighbors determine what is the right place for an atom, the atoms must feel one another out quite extensively in the course of the transformation.

The fact that the melting point of a crystalline solid is very sharp may tempt you to think that it is a fixed property of the material, independent of every surrounding condition other than the temperature. A good way to begin examining whether this is so is to compare the melting point of a solid with the boiling point of a liquid. As you know, the boiling point of a liquid is a variable number; it depends on the pressure over the liquid. "The boiling point," in a table of boiling points, means the boiling point under the pressure of the atmosphere at sea level. A pressure cooker lets the pressure of steam build up above atmospheric pressure, so that the water will boil at a higher temperature and cook things faster. On the other hand, he who climbs from sea level to the top of Pike's Peak must boil his breakfast eggs a little longer than usual.

Now salol would melt at the same temperature at the top of Pike's Peak as it does at home, as nearly as you could tell, and this might lead you to think that a melting point is a fixed number, in contrast to a boiling point. But it is not. If you are not already acquainted with the fact that pressure reduces the melting point of ice, you can easily demonstrate it by an experiment in your refrigerator (Figure 111). Probably this explains why a glacier can flow. The weight of the ice puts a pressure on its lower layers which lets them melt at a lower temperature. Then the glacier moves a little; but since the melting relieves the pressure, the lower layers freeze again, and the process repeats itself.

There is a simple kind of reasoning that makes it clear how pressure changes the melting point of a solid. For example, ice floats on water, and therefore a given amount of water—a given number of molecules—occupies a larger volume when it is solid than when it is

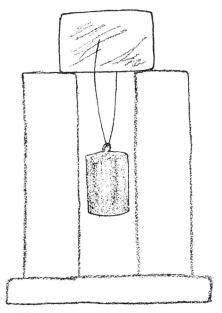

Fig. 111. TO MELT ICE BY PRESSURE, *put an "ice cube" a half-inch thick in a frame so that its thin dimension is horizontal. Hang on the cube a loop of fine wire supporting a weight, and put the assembly in the freezing compartment of a refrigerator. The pressure of the wire on the ice directly beneath it will melt that part of the ice and squeeze the water around the wire. Since that water is no longer under pressure, it freezes again, and the wire travels downward through the ice cube without cutting it in two. In one such experiment a #30 wire carrying a weight of 1¼ pounds traveled ¾ inch in two hours.*

liquid. A pressure applied to ice tries to make the water molecules occupy a smaller volume. They can accommodate this effort by melting. Hence, if the temperature is not too far below the normal freezing temperature, the ice will melt at that lower temperature. In other words, the pressure lowers the melting point. Unlike water, most materials contract when they freeze; therefore, pressure will raise the melting point of the typical solid. Pressure will help to hold the molecules together in solid form at a temperature slightly above the usual melting point.

The effect of pressure on transformation temperatures bears out the same line of reasoning. Tin's low-temperature form occupies the larger volume, and pressure would reduce its transformation temperature below 18° centigrade, just as pressure reduces the melting point of ice below 0 centigrade.

This reasoning also makes it clear why the effect of pressure on a melting point, or on a transformation temperature, is much less than its effect on a boiling point. The difference in the volume occupied by a given number of molecules in a gas and in a liquid is very much greater than the difference for a liquid and a solid, or for two different crystalline arrangements. A small pressure can accomplish much more in restraining the vaporization process than in impeding or assisting the melting and transformation processes. Nevertheless, there will be an infinitesimal change in the melting point of salol if it is taken to the top of Pike's Peak.

CHAPTER XI

The Piezoelectric Effect

In Chapter I we described how salt crystals are built of electrically charged atoms. The chapter pointed out why those "ionic crystals" do not seem to behave as electrically charged objects should; they are electrically neutral as a whole because they contain just as many positive charges as negative charges. Most of the evidence for electric charges in crystals is indirect evidence. For example, in Chapter IX, the fact that the crystals are built of electrically charged units has given a good explanation of why sodium chloride and sodium nitrate cleave easily in certain directions.

But some crystals show a little more directly that they contain electric charges: the charges can be made to produce an electric effect. The phenomenon, called the "piezoelectric effect,"* appears when the crystal is squeezed.

A crystal shows the piezoelectric effect because squeezing the crystal changes its shape. Even if the squeeze is not sufficient to damage the crystal, its shape changes a tiny bit, and it springs back into its original form when the squeeze is removed. In order for a crystal

*The "piezo" part of the word consists of three syllables, pi-e-zo, and the accent falls on the first syllable.

to change shape, its atoms must move. When those atoms are ions, carrying electric charges, the electric charges move with the ions. But a motion of electric charges is an electric current. Hence, when you squeeze a crystal, you can expect to observe an electric current while the crystal is changing its shape.

At first this reasoning may seem to suggest that all crystals should show the piezoelectric effect, but in fact only certain crystals do. A little further reasoning makes it clear why the effect does not occur in many crystals. Every crystal contains an equal number of positive and negative electric charges. When the shape of a crystal changes, both sorts of charges move. If the negative charges and the positive charges move in the same way, the electric current from the moving negative charges cancels out the current from the positive charges, and altogether there is no current at all. A crystal will show the piezoelectric effect only when a squeeze moves the negative charges and positive charges in opposite directions. Consequently, the effect requires special sorts of crystalline orderliness.

Several of the crystals you have grown have the necessary structures. Sodium chlorate and bromate, both of the hydrates of nickel sulfate, lithium trisodium chromate hexahydrate, and Rochelle Salt all show the effect. Indeed, in Rochelle Salt it is unusually large—large enough to see in the simple experiment described in Figure 112 and Plate 38.

It is easy to misinterpret what is going on in this experiment. The sudden pressure applied to the crystal by the hammer seems to be squeezing out electricity, and it is tempting to think that a continuous current would flow if the crystal were squeezed continuously in a vise. In fact, however, current flows only while the squeeze is changing. No current flows so long as the squeeze stays constant.

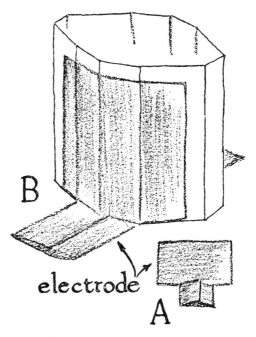

electrode

Fig. 112. To observe the piezoelectric effect, *make two thin electrodes out of aluminum foil, leaving short tabs for electrical leads (A), and stick the electrodes to opposite sides of a Rochelle Salt crystal (B), with a thin film of mineral oil or vaseline. Hold the assembly together by wrapping it in adhesive cellophane tape, and mount it on a cushion of cardboard stuck to a piece of wood. Attach the wire leads from a low power neon glow lamp by wrapping the end of each around a thumb tack and forcing the tack through the tab of an electrode and into the wood. Stick another cardboard cushion on the top of the crystal for an anvil. When you strike the anvil with a hammer, you will see the neon lamp flash.*

In the experiment in Figure 112 a hammer blow applies a rapidly increasing squeeze to the crystal, changing the shape of the crystal suddenly, and producing a small burst of current while the shape changes. When the hammer stops squeezing the crystal, the crystal regains its original shape, and the charges move back to the positions which they originally occupied, giving another small burst of current in the opposite direction. Consequently, although a continuous squeeze would not produce a direct current, a continuous succession of hammer blows would produce an alternating current.

The imaginary two-dimensional ionic crystal of Figure 83 is a good example of an atomic arrangement that can show the piezoelectric effect. The arrangement is shown again in Figure 113A. Each of the white ions

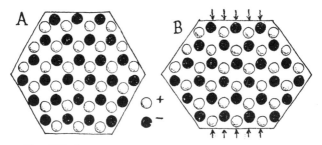

Fig. 113. SQUEEZING THE TWO-DIMENSIONAL CRYSTAL *in the direction of the arrows distorts the atomic arrangement shown at A into that shown at B.*

carries a positive electric charge, and each of the black ions a negative charge. When the crystal is squeezed by pushing the top and bottom toward each other, as in Figure 113B, the ions cannot approach much closer to their nearest neighbors, but they can pry their way toward the open spaces by pushing one another sideways. The result is an orderly distortion of the original arrangement.

Now examine what happens to the electrical charges when this distortion occurs. The best way to look at the problem is to divide the crystal into identical small pieces. Then, since the pieces are all alike, and the squeeze distorts them all in the same way, it is only necessary to find out what happens to one of the pieces. The most obvious way to divide the crystal is by the division into hexagonal pieces shown in Figure 114. The

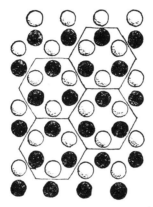

Fig. 114. DIVIDING THE STRUCTURE *of Figure 113 into small imaginary hexagonal units.*

whole crystal can be built of this one sort of pieces, by fitting enough of them together.

When the crystal is not squeezed, the typical piece has the regular hexagonal shape shown in Figure 115A. Notice that its three positive ions are at the corners of an equilateral triangle. Their center of gravity is consequently at the center of the piece, and so is their center of charge. The three negative ions are also on an equilateral triangle, and their center of charge is also at the center of the piece.

The squeeze distorts the piece into the form shown in Figure 115B. The upper ion moves down and the

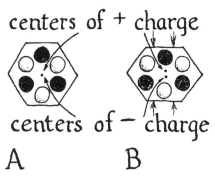

Fig. 115. A UNIT OF THE STRUCTURE of Figure 113. *The centers of positive and negative electric charge coincide (A) when the unit is undistorted. When the unit is squeezed by the forces shown by the little arrows, the centers of the two kinds of charges move in opposite directions (B), producing an electric current toward the top of the page while they move.*

lower ion moves up. The left-hand pair moves to the left, and the right-hand pair to the right, without any change in the vertical separation of the ions within either of the two pairs. The motion of the left-hand and right-hand pairs produces no current, because the current from the motion of the positive charge in each pair is canceled by the current from the motion of its negative companion. Thus, the only uncanceled current comes from the motions of the upper ion and the lower ion.

The upward motion of the positive charge is a tiny electric current upward. But the downward motion of the negative charge is also a tiny current upward, because opposite electric charges moving in opposite directions give electric currents in the same direction. The two currents add together, instead of canceling each other, and hence the whole crystal carries a current upward while the ions are moving into the squeezed positions. When the squeeze is removed, there is a downward current while the ions return.

After examining how this imaginary crystal gives an electric current when it is squeezed, you will have no difficulty understanding why the mineral sphalerite shows the piezoelectric effect. Sphalerite is the mineralogists' name for a crystalline form of zinc sulfide, ZnS. The unit cell of the structure of sphalerite in Figure 99 shows enough of the arrangement of the atoms to make it clear that the sulfur atoms surrounding each zinc atom are placed as if their centers were at the corners of a regular tetrahedron, with the zinc in the middle.

Figure 116 extends the same arrangement to include

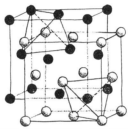

Fig. 116. In the sphalerite structure *each zinc atom is tetrahedrally surrounded by sulfurs, as Figure 99 has shown. When that diagram is extended to include more atoms, it is clear that each sulfur atom is also tetrahedrally surrounded by four zinc atoms. The two sorts of tetrahedra are oppositely cocked.*

a few more of the zinc atoms, and thus to show that they are placed tetrahedrally around each sulfur atom. But the tetrahedra of the zincs around each sulfur are cocked in the opposite direction from the tetrahedra of the sulfurs around each zinc.

Now suppose that a cube of sphalerite is acted upon by forces shown by the arrows in Figure 117B. The cube is squeezed along one diagonal direction and at the same

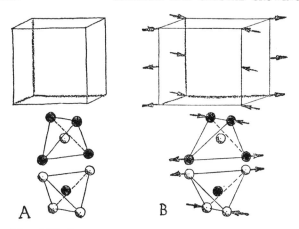

Fig. 117. DISTORTING A CUBE OF SPHALERITE *by the forces shown by arrows changes the regular tetrahedra shown at A into the less symmetrical tetrahedra shown at B. During this distortion the sulfur atoms force the zinc atoms down, and the zinc atoms force the sulfur atoms up. Since the zinc atoms bear a positive charge and the sulfur atoms a negative charge, the motions of both sorts of atoms produce electric currents in the downward direction.*

time it is stretched along another diagonal direction. Then each of the tetrahedra in Figure 117A is distorted (Figure 117B); two of its corners come closer together, and the other two move farther apart. The pair of atoms coming together squeezes the atom at the center of the tetrahedron either up or down, depending on whether that pair is above or below the central atom. The other pair of atoms, the separating pair, opens up a little space for the central atom to move into.

In this way the sulfur atoms push all the zinc atoms down, and the zinc atoms push all the sulfur atoms up, because the two sorts of tetrahedra are oppositely cocked. But the sulfur atoms carry a negative charge and the zinc atoms a positive charge. As in the case of the two-dimensional crystal, the opposite charges moving in opposite directions both produce electric currents in the same direction.

Crystals showing the piezoelectric effect have another interesting property. When an electric voltage is placed across them, they change shape a tiny bit. The effect is so closely related to the piezoelectric effect that it is often called the "converse piezoelectric effect." The "converse" is appropriate for the following reason. In the piezoelectric effect a mechanical cause—a squeeze—produces an electrical effect—a burst of current. In the converse effect an electrical cause—a voltage—produces a mechanical effect—a distortion.

Since the fictitious two-dimensional crystal that was used earlier in this discussion exhibits the piezoelectric effect, it can be expected to exhibit the converse effect also. Figure 118 shows why a mechanical distortion of the crystal occurs when a voltage is applied to it. Again the crystal is imagined as divided into small units. The

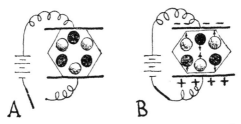

Fig. 118. TO ANALYZE THE CONVERSE *piezoelectric effect, imagine a structural unit (A) of the two-dimensional crystal of Figure 113, placed between electrodes which are connected to a battery through a switch. When the switch is closed (B), the battery moves electric charges to the electrodes, and those charges exert forces on the ions in the crystal, moving them and thus distorting the crystal.*

effect of the voltage on each unit is examined by imagining that the unit is placed between two metal electrodes connected through a switch to a battery, as in Figure 118A. When the switch is closed, a momentary current flows while the battery places its voltage across the little crystal unit. But since the crystal is an electrical insulator, the current cannot flow continuously; it merely flows long enough to put positive charges on one of the electrodes and negative charges on the other, as shown in Figure 118B.

The negative charges on the top plate attract the positively charged white ions and repel the negatively charged black ions. The positive charges on the bottom plate also force the same ions in the same directions. The result is a distortion of the piece, and thus of the whole crystal since it is made of many identical pieces which are all distorted in the same way.

The piezoelectric effect and its converse are so closely related that the magnitude of either in a particular kind of crystal can be calculated after making a measurement of the other effect. Since Rochelle Salt shows an unusually large piezoelectric effect, it also shows an unusually large converse effect. Nevertheless, the converse effect is much too small to be seen without special instruments. Three hundred volts placed across a Rochelle Salt crystal one centimeter thick would change its dimensions only about .001 per cent.

So far you have looked only at the piezoelectric effect and its converse for squeezes and voltages applied in a single direction in the crystal. But, like cleavage, both of these electrical effects depend quite critically on direction. As Chapter I pointed out, such a dependence of the properties of a crystal on direction is one of the most conspicuous pieces of evidence for the repetitive orderly arrangement of atoms.

CHAPTER XII

Some Optical Experiments

"Greatly prized by all men is the diamond, and many are the joys which similar treasures bring, such as precious stones and pearls, though they serve only for decoration and adornment of the finger and the neck; but he who, on the other hand, prefers the knowledge of unusual phenomena to these delights, he will, I hope, have no less joy in a new sort of body, namely a transparent crystal, recently brought to us from Iceland, which perhaps is one of the greatest wonders that nature has produced. I have occupied myself for a long time with this remarkable body and carried out a number of investigations with it, which I gladly publish, since I believe that they can serve lovers of nature, and other interested persons for instruction, or at least pleasure."

So wrote Erasmus Bartholinus,* professor of mathematics and medicine at the University of Copenhagen, in 1669. He referred to calcite, of which beautiful crystals had been discovered not long before near Eskif-

*The quotations from Bartholinus, and later from Huygens, are taken, with minor abbreviation, from *A Source Book in Physics* by W. F. Magie (McGraw-Hill Book Co., Inc., New York, 1935). Figures 119, 120 and 122 are adaptations of Figures 56, 57 and 62 in that source.

jordur in Iceland and promptly called "Iceland spar."
Indeed, this wonder of nature captured the attention of
many other men of science of the time.

Bartholinus continued, "As my investigation of this
crystal proceeded there showed itself a wonderful and
extraordinary phenomenon: objects which are looked at
through the crystal do not show, as in the case of other
transparent bodies, a single refracted image, but they appear double. This discovery and its explanation occupied me for a long time, so that I neglected other things
for it; I recognized that I had come upon a fundamental
question in refraction. In a superficial examination it is
easy to miss seeing this phenomenon, yet it can easily
be exhibited in the following way: we place on a clean
paper any object, for example, a point or something
similar [such as B in Figure 119] and place the lower

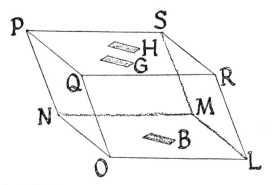

Fig. 119. BARTHOLINUS OBSERVED *that an object B,
viewed through calcite, appears as two images, G and H.*

surface LMNO of the Rhomboid upon it. Then we look
through the upper surface RSPQ [at the object B] by
directing the eye through the whole mass of the prism
RSPQOLMN. . . . If we look through other transparent bodies like glass, water, etc., the image of an object

will appear only once, while in this case we see each of them in a double image on the surface RSPQ [that is B in G and H]. It is to be noticed that the distance between the images H and G which are given from the object B, is greater or smaller according to the size of the prism used; with thin pieces it can hardly be noticed and increases in proportion to the size of the crystal."

Long after this, it was found that crystals of sodium nitrate have the same optical properties as calcite. Today the reason for this correspondence is clear: the atomic arrangements are the same in both. Calcite is composed of calcium carbonate; in sodium nitrate, sodium takes the positions of calcium, and nitrogen the positions of carbon. You can easily duplicate Bartholinus' observations by using sodium nitrate crystals grown by the recipe in Chapter V. Plate 39 shows the sort of thing he saw in calcite.

Bartholinus noticed a peculiarity in the double image: its two parts are somehow different. Said he: "A special property of our crystal is that it gives a double image of one object; another property must be mentioned which is peculiar to it and makes this crystal of special interest among all the minerals. If we look at objects through transparent media, the image remains fixed and immovable in the same position however we move the medium to and fro, and only if the object itself is moved does the image also change its position on the surface of the transparent medium; in this case, on the other hand, we can observe that one of the two images is movable, and this can be established in the following way: [in Figure 120] let the eye and the object A remain at rest; we then turn the prism which is placed on A in such a manner that its lower surface MHL always rests on the table, but so that the edge EM which originally leaned toward G, finally is directed toward F. We can then perceive that one of the images follows the motion of the prism,

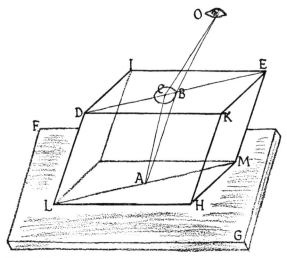

Fig. 120. TURNING A PIECE OF CALCITE *and viewing a pencil dot (A) through it, you see one image (B) of the dot rotate with the calcite while the other image (C) is stationary.*

that is, the image B; for while originally it was seen in the direction of G, it appears, after the prism is rotated, in the direction of F. The image C, however, remains fixed and unmoved. In the future, therefore, the former image will be called the movable one, the latter the immovable or fixed one."

And Bartholinus went on to suggest some of the terminology still used in speaking of this phenomenon. "We now know that an image of an object through two transparent bodies of different nature can only be produced by refraction and that an image requires a refraction; by assuming that the refraction is the cause of the phenomenon in question, it is admissible to draw the conclusion that for the doubled image there should be double refraction. Further, we noticed that the two images produced by our Iceland crystal are not exactly alike, but are distinguished from each other by one of them remaining at rest while the other is movable. We therefore conclude that we can distinguish the two kinds of refraction, and we designate that one which gives us the fixed image as ordinary refraction and the other which gives the movable image, as extraordinary refraction [Figure 121]. The crystal itself we call doubly refract-

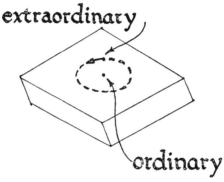

Fig. 121. Bartholinus designated *light rays that turn when the calcite crystal turns as "extraordinary," and light rays that are stationary as "ordinary."*

ing, on account of the extraordinary and peculiar properties of double refraction."

Today we know that Bartholinus somewhat overstated matters when he said that this property was "peculiar to" Iceland spar. Not only sodium nitrate but all crystals in the hexagonal and tetragonal crystal systems (defined in Plate 28) have the property of double refraction, and crystals of the orthorhombic, monoclinic, and triclinic systems have even more complicated optical properties. The peculiarity of calcite and its close allies is that their double refraction is unusually strong; in most crystals only special instruments can detect it.

But it is wholly absent only in crystals belonging to the cubic system.

The Dutch scientist Christiaan Huygens (1629–95) was another who was fascinated by the new crystal. He had a novel theory that light consisted not of a stream of particles, but of waves. In order to show that his theory encompassed all that was known about light, he had to explain Bartholinus' double refraction by it, and he did. But in the course of examining crystals of Iceland spar, he ran across something that puzzled him. In his great *Traité de la lumiere* of 1690 he wrote, "Before closing my treatment of this crystal, I will add an account of a wonderful phenomenon which I discovered after I had written all that has gone before. For, although I have not been able to find the reason for it, I will not refrain from pointing it out, so as to give an opportunity to others to investigate it. It seems necessary to make other assumptions in addition to those that I have made, which, nevertheless, retain all their probability, having been confirmed by so many proofs.

"The phenomenon is this: if we take two blocks of the crystal [Figure 122] and place them one over the other, or hold them separated by some distance, so that the sides of one of them are parallel to those of the other, then a ray of light, such as AB, which is divided into two in the first block, as represented by BD and BC, according to the two refractions, regular and irregular, on passing from one block to the other, so proceeds that each ray passes in it without again dividing into two. The one which has been made by regular refraction, as DG, will make only a regular refraction in GH, and the other CE an irregular refraction in EF. The same thing happens not only with this arrangement, but also in all cases in which the principal sections of the two blocks are in the same plane. It is not necessary that the two surfaces which face each other should be

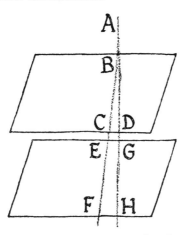

Fig. 122. HUYGENS WAS PUZZLED *by the fact that when a crystal of calcite splits a ray of light into ordinary and extraordinary rays, a second crystal may or (as in this diagram) may not split those rays again, according to the relative orientation of the two crystals.*

parallel. Now it is wonderful that the rays CE and DG, coming from air to the lower crystal, do not divide in the same way that the first ray AB does. We might say that the ray DG, by passing through the upper block has lost some property by which it can move the matter which takes part in the irregular refraction, and that CE has similarly lost that property by which it moves the matter which serves for the regular refraction, but there is one thing which overthrows this hypothesis. When we place the two crystals in such a way that the planes of the principal sections cut each other at right angles, whether the surfaces opposite each other are parallel or not, then the ray which comes from the regular refraction, such as DG, makes only an irregular refraction in the lower block, and on the contrary the ray which comes from the irregular refraction, such as CE, makes now only a regular refraction.

"But in all the other possible positions except those which I have indicated the rays DG, CE divide each into two by refraction in the lower crystal; so that from the single ray AB there are made four rays, sometimes of equal brightness, sometimes so that some are brighter than others, according to the different positions of the crystals, but they do not appear to have more light altogether than the single ray AB." If you have grown two crystals of sodium nitrate, or one large enough to cleave into two, you can set one on top of the other, rotate the top crystal leaving the bottom one at rest, and observe the behavior that puzzled Huygens.

The effort to resolve these puzzles led over the years to an important part of the present understanding of the nature of light. To describe what is happening to the beams of light passing through calcite and sodium nitrate would require a preliminary discussion of light, leading far from the subject of crystals.† Today the light in the two beams emerging from these crystals is called "plane-polarized light."

Some other optical experiments are easily done with the crystals for which Chapter V gives recipes. They are so beautiful that anyone who has grown the crystals will surely wish to perform them.

The first of these experiments provides more evidence that something odd has happened to a light beam when it is split into two beams by a crystal of sodium nitrate. This experiment requires also a crystal of the dark bluish-green copper acetate monohydrate (Chapter V) thin enough to see through. From crystals that are too

---

†In carrying your understanding of optics farther, you can get help from *University Physics* by F. W. Sears and M. Zemansky (Addison-Wesley Publishing Co., Reading, Mass., 2nd Ed., 1955), where Chapters 39, 40, 41, and 47 describe the nature of light and its reflection, refraction, dispersion, and polarization.

thick, thin plates can be cleaved after preliminary exploration for the cleavage plane.

First prepare a sodium nitrate crystal with an optical slit on one of its faces, in the way shown in Figure 123.

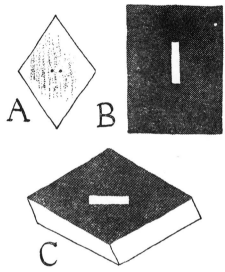

Fig. 123. HOW TO MAKE A DICHROSCOPE. *A—Look through a sodium nitrate crystal at a dot on a piece of paper, and estimate the distance between the two images of the dot. B—Cut a slit in a piece of black paper, making the width of the slit about the same as the apparent distance between the images. C—Stick the paper to the back of the sodium nitrate crystal with rubber cement, with the slit running parallel to the long diagonal of the crystal face.*

Then look through the crystal—with the slit on the back —at light that has first come through the thin slice of copper acetate monohydrate. It is best to hold the slice against the slit, with an electric light or a window behind it. You will see two images of the slit, just as you saw

two images of the dot (Figure 123A). But the two images will differ in an unexpected way: one will appear
light green, the other dark blue.

At first this experiment suggests that the two beams
in the sodium nitrate crystal have different preferences
for colors. But a piece of colored cellophane will yield
only the same color in both beams. Indeed, crystallinity
alone is not sufficient, for a cleavage plate of nickel sulfate hexahydrate will not yield differently colored beams.
The copper acetate monohydrate crystal behaves in an
unusual way toward the light; and the sodium nitrate
crystal, by splitting the light into two beams, reveals that
fact.

If the slice of copper acetate monohydrate is large
enough for you to turn it, still holding it against the
slit on the sodium nitrate, the colors of the two images
will have exchanged places after you have rotated the
slice through a quarter turn. The two crystals are co-
operating with each other in their separate effects on the
light. Indeed, the copper acetate monohydrate crystal
splits the light into two plane-polarized beams, just as
the sodium nitrate crystal does. Instead of sending the
two beams in different directions, it colors one differently from the other. Then the sodium nitrate crystal accepts the two differently colored plane-polarized beams
and sends them in different directions.

There are many other crystals that will co-operate
with calcite and sodium nitrate in a similar way. The
most famous is the mineral tourmaline, a semiprecious
stone long used in jewelry. Not all crystals of tourmaline
will co-operate. But some exhibit the effect in its most
exaggerated form: one of the images is black and the
other is white! In other words, the light is completely
extinguished in one image, and it is partially transmitted in the other.

Polaroid, which has come into use in the last twenty

years, resembles tourmaline in its behavior toward light. The eyepieces from an inexpensive pair of Polaroid sunglasses will provide two sheets of the material for experiment. If you place one of these sheets against the slit at the back of the sodium nitrate crystal, and turn it slowly, the brightnesses of the two images will vary and you can even find a position at which one of the images disappears entirely. Then if you rotate the Polaroid through a quarter turn just as you rotated the slice of copper acetate monohydrate, the other image will disappear.

Polaroid contains crystals that behave like tourmaline. A sheet of it consists of two sheets of plastic with a thin film of little crystals between them. The process by which Polaroid is manufactured turns all the crystals the same way, so that the film is much like a broad, thin, single crystal plate of tourmaline.

You can do an experiment with the Polaroid alone to show that there is something odd about its behavior toward light. Take two sheets of Polaroid—for example the two eyepieces—and hold one behind the other. Turning one slowly and keeping the other fixed, you will see the intensity of the light change, and you can find a position in which the Polaroids transmit no light at all.

There is another interesting optical experiment you can do, with plane-polarized light and a crystal of sodium chlorate (Chapter V), that will give still more insight into the process of crystallization itself. Make a simple "polarimeter": a piece of apparatus which permits you to hold one of the Polaroid eyepieces still and rotate the other and has room between for a crystal of sodium chlorate. A suitable apparatus, with directions for making it out of cardboard, is shown in Plate 40 and Figure 124.

Look first at the light coming through the Polaroids without passing through the crystal. Turn the front Po-

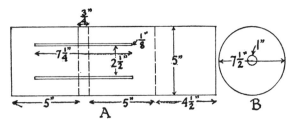

Fig. 124. HOW TO MAKE AND USE THE POLARIMETER *shown in Plate 40. Cut two slots 7¼ inches long, ⅛ inch wide and 2½ inches apart in a cardboard, and score and fold the cardboard along the dotted lines shown in diagram A, sticking its two ends together with adhesive cellophane tape to form a tent-shaped mounting. Prepare two circular discs of cardboard 7½ inches in diameter with holes one inch in diameter cut in their centers. Stick an eyepiece from Polaroid sun-glasses over the hole in each disc, and cement over one disc and its eyepiece a piece of polar coordinate paper with a one-inch hole in its center (diagram B).*

Rest a crystal on the ¾ inch platform between the slots in the mounting, insert the discs in the slots, place a source of light behind the assembly, and rotate the disc carrying the coordinate paper until the light coming past the crystal is extinguished. This gives a zero reading, from which you can rotate the coordinate-bearing disc until the light coming through *the crystal is dark blue.*

laroid until that light blacks out (Figure 125A). Colored light will still come through where the crystal is. Then turn the front Polaroid slowly away from that position. The light passing through the crystal will change color, going through the spectral order. It will never black out; at its darkest it will be a dark blue (Figure 125B).

Using several crystals, with different thicknesses, you will notice that the amount you have to turn the Polaroid in order to reach dark blue will be greater the thicker the crystal. This may become very conspicuous

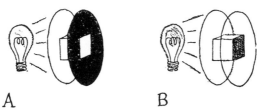

A   B

Fig. 125. SODIUM CHLORATE IN A POLARIMETER.
A—When the Polaroids are placed so that light not pass-
ing through the crystal disappears, you can still see
much of the light which has passed through the crystal.
B—When the second Polaroid is turned so that most of
the light passing through the crystal disappears, light
appears around the crystal.

if you put in two crystals, one behind the other. Whatever is happening to the light is something that increases
with the distance the light travels through the crystal.

But if you put in two crystals, you may have a surprise. You may find a pair of crystals whose effects seem
not to add but to subtract. If you find such a pair, look
at each one separately through the apparatus, noticing
which way you must turn the Polaroid in order to reach
dark blue. You will discover that you must turn it one
way for one crystal and the opposite way for the other.
And you will find, with either of the crystals, that it
makes no difference which pair of faces you choose to
look through. The direction in which you must rotate
the Polaroid is the same for any one crystal, regardless
of the direction the light is taking through the crystal.
You can distinguish the two crystals as a "clockwise"
crystal and a "counterclockwise" crystal, according to
which way you turn the Polaroid to reach dark blue.

This suggests that it might be instructive to look
through the polarimeter at the solutions from which the
chlorate crystals grew—to see whether they also transmit
light when they are between two Polaroids that are

blacking each other out, and if so whether they are clockwise or counterclockwise. It turns out that the solutions do not show the same behavior as the crystals. Polaroids that are blacking out when there is no solution between them black out just as thoroughly when the solution is there. Hence, the ingredients of the crystal—the sodium ions and the chlorate ions—do not have these effects on light when they are in disorder. It is only when they are arranged in order in the crystal that they do. And they can be arranged in two different ways which can be distinguished by their behavior toward the light: a clockwise way and a counterclockwise way.

The fact that any one crystal of sodium chlorate is the same—clockwise or counterclockwise—all through, for light going in any direction, shows that the seed from which the crystal grew must have determined whether it would be of one sort or the other. A clockwise seed imposes its own clockwise structure on all the material growing on it, and a counterclockwise seed similarly imposes its own structure. The ions in the solution have no preference between the clockwise and the counterclockwise form until they encounter the seed. Here is a dramatic instance of the orderliness in a crystal, and of the successful struggle to attain it in the process of crystallization.

It is interesting to speculate on what determines whether a particular seed will be clockwise or counterclockwise when it first forms in some solution. Testing your collection of sodium chlorate seeds in the polarimeter, you almost certainly will find that you have nearly as many of one sort as the other unless you used the dust from a single crystal as "seeds for your seeds." You can guess, therefore, that the choice between these two kinds of orderliness, which must be made by each group of ions coming together to form a seed, is purely accidental.

CHAPTER XIII

Classifying Crystals

The structures of sodium chloride, of copper, and of iron all have the symmetry of the most appropriate building block for their crystals—the high symmetry of a cube. In this respect, if in no other, they are like one another, and they are therefore said to belong to the same crystal class. Of our imaginary two-dimensional crystals, the structures shown in Figures 77 and 79 have a similar property in common. Both have the symmetry of their empty building block, a regular hexagon.

But the two-dimensional structure shown in Figure 83—the fictitious ionic crystal with two different kinds of atoms in it—has a lower symmetry than the regular hexagon, and a "crystal" with that structure would be said to belong to the same crystal system as the other two, but to a different crystal class. Similarly, among real crystals sodium bromate, which belongs to the cubic system because its most appropriate building block is a cube, has not the full symmetry of a cube. The crystal belongs to a different class from that of sodium chloride, copper, and iron. In fact, each of the six systems can be divided into several classes, differing in symmetry.

You have already noticed that the symmetry of a crystal is reflected in many of its physical properties. The piezoelectric effect is a very good instance. A crystal can show a piezoelectric effect only if it lacks a center of symmetry. In each of the six crystal systems there are some symmetry classes with a center of symmetry, and there are others without. If you want to predict whether a particular crystal can show a piezoelectric effect, therefore, you must do more than determine its crystal system; you must take the next step, and determine its crystal class. Conversely, if a crystal exhibits the effect, it must belong to a class having no center of symmetry.

Determining what symmetry a crystal has, and thus assigning the crystal to its proper crystal class, is sometimes quite a difficult experimental problem. The problem would be straightforward, of course, if the arrangement of atoms in the crystal were known. But often an examination with X rays cannot determine the structure unambiguously. The most widely applicable method for determining symmetry is a careful study of the habit of the crystal—the relative sizes of its characteristic faces. The tetrahedral habit of sodium bromate, for example, makes it abundantly clear that the crystal has not the full symmetry of a cube.

When habit fails, other properties must be examined. Sometimes the crystal faces bear helpful markings called "striations." Some crystals can be "etched" in a revealing way. Optical studies of the crystal may avail. The symmetry of the molecules that compose the crystal may put limits on the class to which it can belong. In this chapter you will meet examples of all these kinds of evidence.

Telltale Habits

You may be puzzled by the fact that the internal symmetry of a crystal—the symmetry of its structure—sometimes does and sometimes does not reveal itself in the external symmetry of the crystal—the symmetry of its habit. In order to see how such things can happen, return for a moment to the fictitious two-dimensional ionic crystal whose structure was shown in Figure 83, and consider the "faces" it might have.

As Figure 126 shows, the six faces marked "a" all present the same atomic aspect to the environment in which the crystal is growing. The nutrient material cannot distinguish among these six. When conditions in the environment are the same around the entire crystal, all six faces will grow at the same rate. If they are the only faces appearing, then the "ideal crystal" will have the shape of a regular hexagon, as shown in Figure 83, and will not reveal the lower symmetry of the structure.

But the three little faces marked "b" in Figure 126 may grow slowly enough to appear, as Figure 43 described. These faces differ from the "a's" in two important ways. In the first place, they present a quite different atomic aspect to the environment; they will almost certainly grow at a different rate from the "a" faces.

In the second place, there are only three faces capable of presenting that aspect; they define a triangle, not a hexagon. The three faces that would have to be like the "b" faces, if the crystal had a six-fold axis of symmetry, are marked "c" in Figure 126. Clearly the "c" faces and the "b" faces present different atomic aspects to the nutrient environment.

It would be very improbable to find these two sorts of faces growing at the same rate; the environment would probably behave quite differently toward a face of "white" ions and a face of "black" ions. Even if you visualize coating the two sorts of faces with the same sort of ions, you will reach the same conclusion. There is a different disposition of the "black" ions directly be-

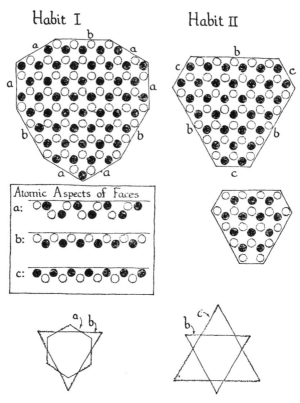

Habit I

Habit II

Atomic Aspects of Faces
a:
b:
c:

Fig. 126. Two habits *of a fictive two-dimensional crystal. The* a *faces, which alone would give the "ideal" crystal the hexagonal appearance of Figure 83, are here accompanied by* b *faces and yield Habit I: a combination of a hexagon and a triangle, which reveals the true symmetry of the structure. If the* c *faces grow slowly enough, the* a *faces may disappear altogether, to yield the different Habit II: a combination of two unequally developed triangles, which again reveals the symmetry of the structure.*

neath the "white" in the coatings, and the nutrient environment will respond to that difference also.

In summary, if the "a" faces and the "b" faces are the slowest growing faces, the crystal may show Habit I—a hexagon, three of whose corners are cut off by a triangle. The symmetry of the resulting figure contains only those operations common to the two simpler figures, and here that is the symmetry of the triangle. If the "b" faces and the "c" faces are the slowest growing faces, the crystal may show Habit II—two triangles of different sizes cutting off each other's corners. Again the symmetry is that of a triangle.

Similar situations often arise among real crystals; sodium chlorate and sodium bromate are good examples. The cubical habit of sodium chlorate is analogous to the habit of the two-dimensional crystal shown in Figure 83. Those habits have higher symmetries than the structures. The faces that appear belong to special sets forming highly symmetrical figures.

Adding a little borax to the growing solution for sodium chlorate reduces the relative growth rate of the tetrahedron faces to the point where they appear on the crystal, cutting off four of the corners of the cube, as shown by one of the drawings in Figure 127. This be-

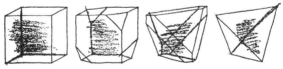

Fig. 127. HABITS OF SODIUM CHLORATE *modified by borax. With increasing amounts of borax in the solution, the tetrahedron faces become more and more prominent.*

havior is similar to the development of Habit I (Figure 126) on the two-dimensional crystal.

On the other hand, the habit of sodium bromate is

PROBLEM 9

1. The squeeze used in Chapter XI in analyzing the piezoelectric effect in the fictitious two-dimensional ionic crystal (Figure 113) was perpendicular to the "b" and "c" faces (Figure 126). Would you expect to observe a piezoelectric effect in that substance if the squeeze were applied to the "a" faces? If so, in what direction would electric current flow as you applied the squeeze?

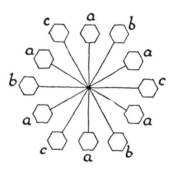

2. The diagram above shows directions perpendicular to the "a," "b," and "c" faces of the two-dimensional crystal. From the combined results for the directions of current flow on application of a squeeze, obtained in Chapter XI and in Part 1 of this problem, indicate by an arrow, in each of the hexagons on the diagram, the direction of current flow when the squeeze is applied along the line to which the hexagon belongs. Compare the symmetry of the resulting diagram with the symmetry of the structure.

analogous to Habit II of the two-dimensional crystal. Usually the crystal forms a tetrahedron whose corners are cut off by another tetrahedron (Figure 128). If the

Fig. 128. THE HABIT OF SODIUM BROMATE *shows the faces of two unequally developed tetrahedra.*

two tetrahedra had the same size, they would combine to produce an octahedron. But the structure has a lower symmetry than an octahedron; the two tetrahedra grow at different rates, as do the two triangles which form Habit II in Figure 126.

It is reasonable that sodium chlorate and sodium bromate should both exhibit a tetrahedral habit. X-ray investigation shows that they both have the same crystal structure, in which chlorine or bromine alternatively plays the same role. But it is unexpected that when they grow from pure solutions, the cube faces grow more slowly than the tetrahedron faces on the chlorate, and the relative growth rates are reversed on the bromate. Together they show dramatically that the relative growth rates on crystal faces can be very sensitive to small influences.

The habit of strontium formate dihydrate is a quite reliable symptom of its crystal class. Occasionally it crystallizes as shown in Figure 129A, with three 2-fold axes, three reflection planes, and a center of symmetry. But much more frequently its crystals also bear the four little faces marked by arrows in Figure 129B. These little faces show that the structure cannot have the three reflection planes or the center of symmetry. They place it unquestionably in that class of the orthorhombic system for which the only symmetry operations are three mutually perpendicular two-fold axes of symmetry.

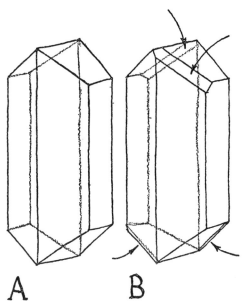

Fig. 129. STRONTIUM FORMATE DIHYDRATE *often shows a habit (A) more symmetrical than the true symmetry of its structure. More often, however, four additional little faces (B) reveal the true symmetry.*

On the other hand, Rochelle Salt, even though it often shows little slanting faces at its ends, almost never exhibits faces that reveal its true symmetry. As you have seen, you can change the habit of Rochelle Salt greatly by adding copper ions to the growing solution. But those habit changes never involve the appearance of new faces which might assist the diagnosis of its symmetry. Its apparent crystal class has the same symmetry as that of Figure 129A; whereas its true crystal class has the symmetry of Figure 129B. Before you finish this chapter, you will see the evidence and reasoning that lead to the correct classification.

In all the cases so far discussed, when habit has failed

to reveal the true symmetry of a crystal, it has suggested too high a symmetry. The discussion of the fictitious two-dimensional crystal has shown why. The slowest growing faces, dominating the crystal shape, may happen to be a set forming a figure consistent with several crystal classes, and therefore in particular with a class having high symmetry.

Much more rarely a crystal may have a habit suggesting a symmetry lower than its true symmetry. Nobody knows why this occurs. An example is potassium chloride, known to mineralogists as sylvine. Repeated observation of habit led early crystallographers to assign potassium chloride to a cubic class having no plane of symmetry. But X-ray investigations have shown it to have the structure of sodium chloride (Figure 97), and have thus placed it in the class that has the full symmetry of a cube.

Striations and the Thirty-two Crystal Classes

A common cubic crystal having a lower symmetry than the cube is the mineral pyrite, sometimes called "fool's gold" because its color occasionally deceived the early prospectors. Pyrite often shows its symmetry in another way than by its habit. The faces on a naturally occurring cube of pyrite may have markings called "striations" or "striae"—little ridges and channels running along the faces parallel to the edges of the cube. As Figure 130 shows, these striations take different directions on the different faces of the cube.

Just as clearly as the additional little faces on strontium formate dihydrate, shown in Figure 129, the striations on pyrite indicate the symmetry of its structure. Notice, for example, that a cube bearing such markings no longer has axes of four-fold symmetry; the four-fold axes of the unmarked cube are degraded to two-fold axes

Fig. 130. STRIATIONS ON A PYRITE CRYSTAL *show that the structure does not have all the symmetry operations of a cube.*

on the marked cube. The three planes of reflection parallel to the cube faces remain, but the six planes through opposite edges disappear. The marked cube still retains four 3-fold axes and a center of symmetry.

Not many crystals are as obliging as pyrite in providing striations to reveal their symmetry. One convenient way to study different kinds of symmetry, however, is to *imagine* crystals having the simple shapes of their appropriate building blocks but bearing striations. It turns out that a cubic crystal—a crystal whose appropriate building block is a cube—can have any one of five different combinations of the symmetry operations discussed in Chapter VII. Each combination defines a crystal class; in other words, there are five cubic crystal classes. Figure 131 shows you a way to make simple models with striations, illustrating the symmetries which define the five cubic classes.

At the low end of the symmetry scale, a crystal belonging to the triclinic system, defined in Plate 28, can have either of two kinds of symmetry. It can have a center of symmetry, as do the crystals shown in Figure 71, or it can have no symmetry at all. Altogether there are thirty-two, and only thirty-two, kinds of symmetry that crystals can have. Each of these thirty-two crystal classes belongs to just one of the crystal systems. The classes are a way of subdividing the systems.

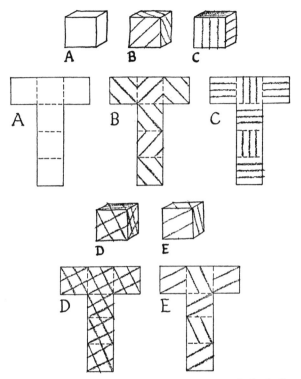

Fig. 131. MODELS WITH STRIATIONS *can easily be built to illustrate the symmetries of the five cubic crystal classes.*

There are many ways of becoming better acquainted with these different symmetry classes. One of the best ways is to examine models which belong to different symmetry classes, turning them in your hands and verifying their symmetries by examining them carefully.

It is easy to make the cardboard models shown in the Appendix, to illustrate the symmetries of the thirty-two crystal classes, in the same way that the models shown in

Figure 131 are made. Each represents a hypothetical crystal whose faces take the simple shape of the building block appropriate to its crystal system (Plate 28). Some of the faces on the crystal bear striations like those on crystals of pyrite. Each of the six models without striations represents the crystal class with the highest symmetry in the system to which it belongs. The striations on the other models of the same shape lower the symmetry in various ways.

In examining a model, it is instructive to assure yourself of two things: (1) that it has *all* the symmetry operations of its class, and (2) that it does *not* have the remaining symmetry operations of the polyhedron of the same shape without striations.

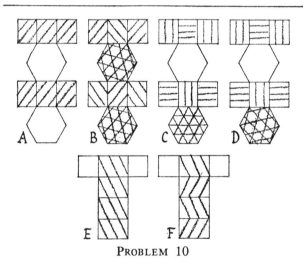

PROBLEM 10

Assign each of the six striated models which can be constructed from these patterns to its appropriate crystal class, with the aid of the table of crystal classes in the Appendix.

Etch Figures

Another way to learn about the symmetry of a crystal is to examine "etch figures" on its surfaces. When a crystal begins to dissolve, the process of dissolution often begins at a few isolated points on the surfaces. Dissolution then spreads from those points, and the manner in which it spreads is obedient to the symmetry of the crystal. Since the velocity of growth of a crystal depends on direction, determining the habit and often revealing the symmetry, it is not unexpected that the reverse process, dissolution, should show some similarity.

Etch figures are most useful in diagnosing symmetry when they take the form of little shallow pits, with pyramidal shapes that can be seen under a low-power magnifying glass. The shapes of the pits and the angles at which their edges stand to the edges of the face can then be used as symptoms of the symmetry in much the same way as the striations on pyrite.

Revealing etch figures are best produced on a crystal if you use a solvent that attacks the surface slowly, so that the crystal can be dried before dissolution has proceeded so far as to obliterate the figures. Dissolution is usually a much more rapid process than crystallization, but its velocity declines with approach to saturation. For this reason, either a nearly saturated solution or a solvent in which the crystal is only slightly soluble makes the most promising etching fluid.

Two of the crystals for which we have given recipes provide interesting etch pits easily—calcium copper acetate hexahydrate and nickel sulfate hexahydrate. Both belong to the tetragonal crystal system, and both have directions of "perfect cleavage." The cleavages enable you to produce fresh flat surfaces with a known direction. Etch figures can also be developed on natural growth faces, but those faces usually bear the marks of previous handling.

Cleavage of calcium copper acetate provides surfaces parallel to the four-fold axis of symmetry. If you dip a cleavage plate in water for a few seconds, then quickly dry it on a tissue, you will find that it has been etched by the water in long streaks parallel to the four-fold axis, reminiscent of the striations on pyrite.

You can obtain more interesting figures, and at the same time see how sensitive etching can be to small influences, if you use a different solvent. Hold a cleavage plate of calcium copper acetate with tweezers, swish it gently around in 25 cc. of glacial acetic acid for ten seconds, and then dry it. Looking at it through a low-power magnifying glass, you will see shallow four-sided pyramidal etch pits on the surface, resembling Figure 132A.

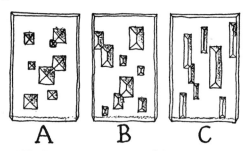

Fig. 132. ETCH FIGURES *on calcium copper acetate hexahydrate. Acetic acid gives square etch pits (A). Small additions of water to the acetic acid elongate the etch pits produced (B and C) along the axis of four-fold symmetry.*

These etch pits are almost square; in fact if they were the only evidence for symmetry, you might think there was a four-fold axis of symmetry *perpendicular* to the cleavage face. But now add one cc. of water to the 25 cc. of glacial acetic acid, and etch a fresh cleavage plate in the same way. This time the etch pits will be somewhat elongated along the direction of the true four-fold axis, as shown in Figure 132B. If you add another cubic centimeter of water to the etching fluid, so that there are two in all, and etch a third plate, you will produce pits still further elongated (Figure 132C). You can see how with pure water they become elongated into streaks.

In nickel sulfate hexahydrate the cleavage planes are perpendicular to the four-fold axis of symmetry of the crystal, and repeated cleavage produces plates whose major surfaces lie in those planes. A slightly unsaturated solution of nickel sulfate in water makes a good etching fluid. Mix 5 cc. of water into 25 cc. of saturated solution, and swish the plate for twenty seconds. The etch figures will resemble those produced by pure acetic acid on calcium copper acetate: four-sided pyramidal pits such as those in Figure 133A. On nickel sulfate they will be so large and so nicely separated that you can see them with the unaided eye.

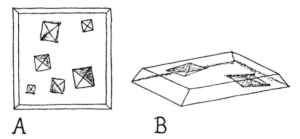

A   B

Fig. 133. ETCH FIGURES *on nickel sulfate hexahydrate, made by an unsaturated solution, are turned a little with respect to the edges of the crystal. A comparison of the figures on opposite sides of a cleavage plate shows that the only symmetry operations for the crystal are a four-fold axis and four 2-fold axes perpendicular to the four-fold axis.*

The square pyramidal shape of the pits simply confirms what you already knew: the plates are perpendicular to the axis of four-fold symmetry. But looking at the pits carefully, you will find that their edges are not quite parallel to the natural faces forming the edges of the cleavage plate. The pits are turned slightly about the four-fold axis. This shows that the crystal does not have planes of symmetry containing the four-fold axis. The external form of the crystal, shown in Figure 69, has those planes of symmetry, and therefore the crystal cannot have all the symmetry operations of its external form.

If you look at etch pits on both sides of the same cleavage plate, you will see that they are turned in the way shown in Figure 133B. This disposition of etch pits shows that the crystal has four 2-fold axes perpendicular to the four-fold axis, and no other element of symmetry. In other words, crystals of nickel sulfate hexahydrate belong to the "Tetragonal 4" class described in the Appendix.

Screw-like Behavior in a Crystal

Already you have noticed that the symmetry of sodium bromate must be lower than the symmetry of a cube, even though it belongs to the cubic crystal system. From the tetrahedral habit of sodium bromate, and from the fact that you can induce the same habit in sodium chlorate by adding borax to the growing solution, you will be tempted to assume that their structures have the symmetry of a tetrahedron. In other words, you will be tempted to assign them to the "Cubic 2" crystal class.

Actually their structural symmetry is even lower than that. They belong to the class listed as "Cubic 5"; their structure lacks the six planes of symmetry of the tetrahedron. On rare occasions, when it grows very slowly from a pure solution, some of the edges of a cube of sodium chlorate will be cut off by little faces which show that the crystal might not have those planes of symmetry. But our knowledge of the low symmetry of sodium chlorate does not rest on that rather risky testimony. An optical observation, which you have already made, gives unquestionable evidence of it.

Using the polarimeter (Figure 124 and Plate 40) with a crystal of sodium chlorate, you have seen that the amount by which the Polaroid must be turned in order to reach dark blue is greater the longer the path of the light through the crystal. You can easily make that observation quantitative, even with a polarimeter as crude as yours. Plotting the rotation of the Polaroid horizontally and the thickness of the crystal vertically, as in Figure 134, you will find that the rotation is directly proportional to the thickness.

Thus as plane-polarized light progresses through the crystal, some rotational feature is involved. Whatever it is, it behaves like a wood screw turning as it advances into the wood. But since plane-polarized light exhibits this behavior only while passing through the crystal, it is the crystal, not the light, that possesses the screw-like property. Indeed, the light may behave either like a right-hand screw or a left-hand screw, depending on whether it is passing through a clockwise crystal or a counterclockwise crystal.

Since any sodium chlorate crystal possesses a screw-like property, its structure must have a symmetry consistent with that property. In particular, it cannot have any planes of reflection symmetry. If the structure had a plane of symmetry, every right-hand screw in it could be paired off with a left-hand screw, and the two would annul each other's effects on the light.

Now since sodium chlorate can be made to grow in tetrahedra, the symmetry of the crystal cannot be higher

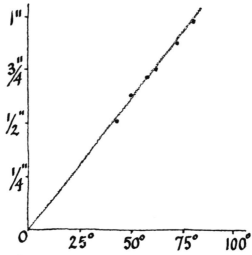

Fig. 134. MEASUREMENTS ON SODIUM CHLORATE *in a polarimeter. A point shows a measurement on a sodium chlorate crystal, in a polarimeter of the type shown in Figure 124. The vertical position of the point is the thickness of crystal through which the light passed. The horizontal position of the point is the angle through which the second Polaroid was turned to attain maximum extinction of that light. The data show that the effect of the crystal on the light is proportional to the distance the light travels in the crystal.*

than that of a tetrahedron: four 3-fold axes, three 2-fold axes, and six planes of reflection. But the optical observation makes it clear that the crystal must lack the planes of symmetry. Hence, it can have only those symmetry operations that remain after you subtract the planes of reflection from the symmetry operations of a tetrahedron. In short, sodium chlorate must belong to the least symmetrical of the five cubic crystal classes: "Cubic 5," which has only four 3-fold axes and three 2-fold axes.

Asymmetric Molecules

There are a great many crystals whose structures are quite complex but whose crystal class can be determined, nevertheless, by experiments and reasoning similar to those which you have just used for sodium chlorate. Rochelle Salt, which you used to demonstrate the piezo-electric effect, is an example. The important difference is that the reasoning is applied directly to the *crystal* of sodium chlorate, but to the *ions which compose the crystal* of Rochelle Salt.

In sodium chlorate the sodium ions and the chlorate ions do not rotate the plane of plane-polarized light until they take up an orderly arrangement in the crystal. As proof of this, Chapter XII pointed out that a solution of sodium chlorate does not affect the light in this way. And that is where the behavior of Rochelle Salt differs; a solution of Rochelle Salt *does* affect the light.

Like a crystal of Rochelle Salt, a solution of it contains potassium ions, sodium ions, and tartrate ions. Solutions of sodium chlorate and of alum, which contain sodium ions and potassium ions, respectively, do not affect light in this way. It must be the tartrate ions that are doing the trick. In fact, a solution of tartaric acid will do the same sort of thing.

You can reason about the tartrate *ions* in the same way that you reasoned in the last section of this chapter about the sodium chlorate *crystal*. The tartrate ions, like the sodium chlorate crystal, must have a screw-like property. And that screw-like property must be either right-handed in them all or left-handed in them all. If it were not—if some ions were like right-handed screws and some like left-handed screws—the ions would cancel one another's effects on the light.

Now consider what happens when many ions, all with

the same screw-like property, are put together into a crystal. The crystal must acquire their screw-like property, at least in part. In particular a screw lacks any plane of symmetry, and a center of symmetry, and therefore the crystal must lack them also. If the crystal had planes of symmetry, for example, every right-handed screw could be paired off with a left-handed screw, its reflection in a plane of symmetry.

This is the argument permitting you to fix the crystal class to which Rochelle Salt must belong. From measurements of the angles between its faces you know the shape of its building block; you know that the crystal belongs to the orthorhombic crystal system (Plate 28). But you also know, from the table of crystal classes, that there are only three crystal classes—three consistent bundles of symmetry operations—belonging to the orthorhombic crystal system. Only one of those bundles includes no plane of symmetry. Hence the crystal must belong to the class listed as "Orthorhombic 3."

Your study of symmetry operations enables you to put this argument in another way, which is more elegant. Since the ions have a screw-like property, they must lack all symmetry operations that are "nonperformable." When you assemble the ions into a crystal, you move them and turn them, but surely the only operations you perform on them are performable, because you actually *perform* them. Performable operations, carried out on objects possessing only performable symmetry operations, cannot construct something that has nonperformable symmetry operations.

You may feel that this is an unnecessarily roundabout way of determining the crystal class appropriate to Rochelle Salt. Since the crystal contains screw-like ions contributing their screw-like property to the crystal, why not look at the crystal itself between sheets of Polaroid, just as we looked at the crystal of sodium chlorate, and reach our conclusion directly? If you try this, you will learn at least a part of the answer. You will be able to find *no* position of the Polaroid sheets in which they extinguish the light when the crystal is between them. You were able to make a useful observation of the crystal of sodium chlorate because it belongs to the cubic system, and has the especially simple optical properties shared by all crystals belonging to that system. The optical properties of Rochelle Salt are complicated by the fact that it is doubly refracting, unlike cubic crystals.

Ordinary sugar also exhibits the same sort of screwlike property. A solution of sugar in water will behave in the same way as a solution of Rochelle Salt toward light passed through a sheet of Polaroid. In fact, if you measure the angle through which you must turn the second Polaroid sheet in order to extinguish light passed through either of these solutions, you will find that it is directly proportional to the path the light takes through the solution. In other words, the size of the effect depends directly on how many of the screw-like ions or molecules the light meets in the course of its travels. And again the symmetry of a sugar crystal obeys the dictates of its screw-like molecules. Sugar crystals belong to the "Monoclinic 3" crystal class, the only class of the monoclinic system lacking a plane of symmetry.

Since the optical effect of a solution of Rochelle Salt or of sugar depends on the number of screw-like molecules that the light meets, it will depend not only on the length of the path of the light through the solution but also on the concentration of the solution. Light passing a given distance through a solution will be more strongly affected by a concentrated solution than by a dilute solution, because the more concentrated solution presents more molecules to the light. This fact is used in an instrument called a "saccharimeter," which measures the concentrations of solutions of sugar, at various stages in its manufacture, by measuring the optical effect of the solutions.

Notice that both of the substances, sugar and Rochelle Salt, which have this effect on light are derived from living things. Sugar is extracted from sugar cane. The tartaric acid used to prepare Rochelle Salt is a by-product of the making of wine from grapes. Indeed, a great many other products derived from living things have this property.

Of course, not all do. The, alcohol in wine does not. The acetic acid, in the vinegar made by letting wine ferment longer, does not. But the arresting fact is that *no* purely synthetic chemical product has this effect on light. Unless some "life process" takes part, either in making the starting material for the chemical operations or as one of the processes deliberately used in manufacture, the product will be as inert as water or glass between two sheets of Polaroid.

You may protest that your crystals of sodium chlorate have this effect on light, whereas their solution does not, and no life process intervenes between the solution and the crystals growing from it. But remember that you find just as many left-handed as right-handed seed crystals growing from the solution when you let those seeds deposit spontaneously. The only way you can grow crystals that are all right-handed, for example, is by picking out right-handed seeds and allowing them to grow alone. In other words, the "life process" which intervenes is *you*—the choice you make among your collection of seeds.

In a similar way a chemist may make tartaric acid by chemical operations on other materials that have no screw-like property. When he has finished his synthesis, his tartaric acid will contain as many right-handed as left-handed molecules, just as your seeds of sodium chlorate are evenly distributed among the two kinds. No chemical test can distinguish the two kinds of molecules, and they will annul one another's effects on the light. No eye is sharp enough, no tweezers fine enough, to pick out the right-handed molecules from the left-handed molecules and put them into a separate bottle. The chemist must set about separating them in some less direct way. And that way must use some life process, or some product of a life process.

For example, more than a hundred years ago Louis Pasteur found that certain moulds, growing in a solution of ammonium tartrate containing both kinds of molecules will destroy the right-handed molecules and leave the left-handed molecules. The enormous molecules of the proteins which form the flesh of animals are made of smaller molecular units, most of them capable of existing in both right-handed and left-handed forms. When rabbits are fed mixtures of the two forms of these units, they assimilate the form from which they can build a useful protein, and excrete the other.

It is easy to guess that the enzymes accomplishing these selections must themselves be capable of existing in right-handed and left-handed forms, and that only one or the other of these forms is actually found in nature. One chemist has graphically described their action: "The enzyme fits the substrate as the key fits the lock." Another instance of this "fitting" is the drug adrenaline, often used in surgery to reduce bleeding by contracting the blood vessels. Adrenaline is capable of existing in right-handed and left-handed forms, but only one of the forms is useful; the other has little effect upon blood vessels.

Today there is nothing really mysterious about the chemical structure of molecules that can take both right-handed and left-handed forms. Their screw-like property need be no more than the lack of a plane of symmetry and a center of symmetry. They are molecules whose atoms are bonded together in an arrangement that leaves them without those symmetry operations. Most of these molecules contain at least one atom of carbon, and it is quite easy to see how a molecule containing an atom of carbon can lack a plane of symmetry.

First, recall how the atoms of carbon are arranged in a crystal of diamond (Figure 96). Each is tightly bonded to four others arranged at the corners of a tetrahedron. This behavior of a carbon atom is quite habitual. For example, in each molecule of methane gas (CH_4), a carbon atom is surrounded by four hydrogen atoms in the same tetrahedral arrangement. Carbon tetrachloride, the familiar cleaning fluid, is made of molecules containing one carbon atom apiece, tetrahedrally bonded to four chlorine atoms.

As Figure 135A shows, the symmetry of a molecule of

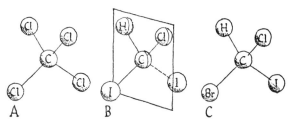

Fig. 135. CARBON FORMS "TETRAHEDRAL" BONDS. *When the other atoms are alike (A), each molecule has the symmetry of a tetrahedron. When only two of the atoms bonded to a carbon atom are alike (B), the molecule still has a plane of symmetry. When all four atoms bonded to the carbon atom are different (C), the molecule is asymmetric.*

methane or of carbon tetrachloride is the same as the symmetry of a tetrahedron. In particular both molecules have the six planes of symmetry of a tetrahedron. But if you make a molecule in which the four atoms bonded to the carbon atom are not all of the same species, the symmetry of the molecule is less than that of a tetrahedron. Such a molecule as that of di-iodo-chloro-methane ($CHClI_2$) in Figure 135B has only one plane of symmetry, the plane through the carbon, hydrogen, and chlorine atoms. Finally, if you make all the atoms in the molecule different—if you make iodo-bromo-chloromethane, for example—the molecule no longer has any plane of symmetry (Figure 135C). In fact, it has no symmetry at all.

And now you recognize that you are talking about a molecule that is like a left hand. Turn it as you will, you cannot transform it into a right hand. But you can imagine a molecule that is just like the left-handed molecule except that it is a right-handed molecule. Figure 136 shows the two molecules side by side. One could

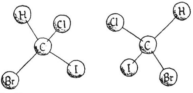

Fig. 136. RIGHT-HANDED AND LEFT-HANDED *forms of the molecule in Figure 135C. Each is like the other form reflected in a plane.*

be obtained from the other only by reflecting it, not by turning it. Any chemical operation by which you make these molecules will make just as many of one kind as the other, and then you are faced with both kinds mixed up together. Almost always, when the two kinds of molecules are mixed together, they form crystals in which the two kinds are still mixed together, and the crystal has a plane or a center of symmetry. This mixture is like that of potassium sulfate and aluminum sulfate in alum; equal numbers of each kind of molecules cooperate to form the orderly structure.

In rare cases the two kinds of molecules will crystallize out separately. Then a crystal made from either kind of molecules alone will lack a plane of symmetry. A crystal of one kind will look like the mirror image of a crystal of the other kind, just as a right hand looks like the mirror image of a left hand. And a visible life process—you and your tweezers—can serve instead of an invisible one, to pick the two kinds of molecules from each other.

In both the tartrate ion and the sugar molecule there is a similar lack of symmetry, which comes from a similar source: the arrangement of groups of atoms of four different sorts tetrahedrally about a single carbon atom. Figure 137 shows the right-handed and left-handed

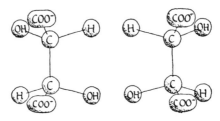

Fig. 137. RIGHT-HANDED AND LEFT-HANDED TARTRATE *ions. Even though the two halves of each molecule can rotate freely about the axis of the carbon-carbon bond when the ion is in solution, the two forms remain distinguishable, and one looks like the reflection of the other.*

tartrate ions. The sugar molecule is more complicated. The tartrate ion does not wholly lack symmetry as the molecule in Figure 136 does; it has an axis of two-fold symmetry. Nevertheless, it has the all-important lacks of a plane of symmetry and a center of symmetry.

What started the right-handedness and left-handedness now so vital to living things? If some life process must accompany the separation of right-handed from left-handed materials, how could life begin in the first place? If everything now right-handed were suddenly made left-handed and everything left-handed were made right-handed, would life proceed without interruption and without change? These are deeply studied problems. Some good guesses have been made, but they have not produced the full answer. The guesses have come—and perhaps the full answer will come—from examining the symmetries of the forces acting on matter, along with the symmetries of the matter on which they act.

Crystals furnish one of the most beautiful examples of order and symmetry. Thinking further, you will find that order and symmetry have very far-reaching applications in all the natural sciences. Today physicists think of the symmetries of the fundamental particles of which the universe is made, of the fields of force in which those particles find themselves, and of the mathematical equations which describe the resulting behavior of the world. Chemists think of the symmetries of the molecules, simple and complicated, made of those particles. And biologists think of those symmetries—that orderliness—in their effort to understand the origin and behavior of life itself.

APPENDICES

SOURCES OF MATERIALS

A few of the chemicals suggested for growing crystals in this book are easily available: they have some other use in day-to-day life. A little ingenuity is needed to obtain others. High school science teachers may be helpful, or a neighborhood druggist might be willing to accept special orders; tell these people what you want the chemicals for. Edmund Scientific, Barrington, New Jersey 08007, has for years offered a collection of the chemicals suggested in this book for crystal growing. They have a mail order catalog.

DRUGSTORE

silver nitrate	potassium ferricyanide
nitric acid	acetic acid
salol (phenyl salicylate)	formic acid
boric acid	borax
Rochelle Salt	mineral oil
sodium hydroxide	rubbing alcohol
sodium nitrate	"Polaroid" sunglasses
	thermometer

While drugstores do not carry the following items as part of their regular stock, they can order them for you.

glass cylinder, graduated in cc.	sodium dichromate
	copper sulfate
potassium aluminum alum	chrome alum
nickel sulfate	copper acetate
sodium bromate	calcium oxide
sodium chlorate	lithium carbonate
strontium carbonate	

CAMERA SHOP

thermometer	chrome alum
glass cylinder, graduated in cc.	silver nitrate
	potassium ferricyanide
potassium aluminum alum	acetic acid

HARDWARE STORE

hacksaw blade
nuts, bolts, washers
thermometer
measuring cup
Mason jars

batteries
1/25-watt neon glow lamp
      or
neon electrical test light
copper sulfate

STATIONERY SHOP

polar graph paper
hand glass

TOY STORE

Plasticine

CONVERSION OF UNITS

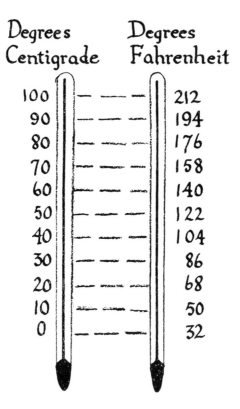

The freezing point of water is 0° centigrade, 32° Fahrenheit. The boiling point of water is 100° centigrade, 212° Fahrenheit. A change of temperature of 10° centigrade is the same as a change of 18° Fahrenheit.

Weight

1000 grams = 2.205 pounds
1 pound = 453.6 grams

Volume of Fluid

100 cubic centimeters (cc.) = 3.38 fluid ounces
1 fluid ounce = 1/20 pint = 29.57 cubic centimeters

1 cubic centimeter of water weighs 1 gram.
A sixpence weighs approximately 3 grams.
A halfpenny weighs approximately $5\frac{3}{4}$ grams.

HISTORICAL OUTLINE

1597 Libavius—Identification of salts in mineral waters by crystal shape.

1670 Bartholinus—Cleavage and double refraction (of calcite).

1690 Huygens—Polarization of light.

1772 Romé de Lisle ⎱

1782 Haüy ⎰ Accurate measurement of crystal angles and their characteristic values for substances, constructibility of crystals from elementary building blocks, the six crystal systems.

1803 Black—Latent heat of melting (posthumous publication).

1808 Widmannstätten—Etch figures (on meteoric iron).

1811 Arago—Rotation of plane of polarization of light in quartz.

1818 Brewster—Relation between optical properties and symmetry.

1819 Mitscherlich—Isomorphism and polymorphism.

1830 Hessel—The thirty-two crystal classes.

1880 Curie—Piezoelectric effect.

1912 von Laue—X-ray diffraction by crystals.

MAKING A SPECTROSCOPE

Since crystals of sodium bromate grow in regular tetrahedra, the angle between any pair of major faces is an acute angle: 70° 32′. Such a crystal is therefore usable as an optical prism; it enables you to make a spectroscope for which you grow your own prism, instead of buying a glass prism.

If you have never seen light spread out into a spectrum, it is worth your while going to the small trouble of cutting a half-inch vertical hole in each end of a shoe box, and forming a very narrow slit over each hole with friction tape. Put a light bulb outside the box at one end, and mount a crystal of sodium bromate on a lump of Plasticine outside the box at the other end. Adjust the crystal so that the ribbon of light transmitted through the slits strikes one face at an angle and emerges from the other face. Holding a piece of white paper a foot or two away, you will find that the yellow light has been bent more than the red light and less than the blue light.

Spectroscopes have been important tools in exploring the electronic structures of atoms. You can begin to see the reason for this if you use other sources of light than an electric light bulb for your spectroscope. Buy a small alcohol lamp (or make one such as Plate 41 shows), and support a little common salt in the eye of a needle, about a half inch above the wick, in the flame of the lamp. The brilliant yellow light is characteristically emitted by excited atoms of sodium in the slowly vaporizing salt. Instead of the entire spectrum, which you get from white light, the spectroscope will show you a single yellow line of light.

Here is the origin of the name "spectral line," which you have probably heard. Almost all spectroscopes use slits to produce a beam of light that does not spread out. When that beam has been split into its components by a prism or a ruled grating, each component appears in an image of the slit: a line of light.

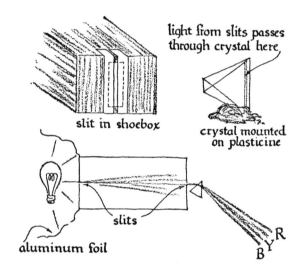

light from slits passes through crystal here

slit in shoebox

crystal mounted on plasticine

slits

aluminum foil

R
Y
B

Using the same principles, you can construct more refined instruments than the shoe-box spectroscope. The photograph of the wooden spectroscope (Plate 42) suggests the next step: a more rigid mounting, narrower slits, more accurate alignment of the crystal, and means for measuring the angles at which the various spectral lines are refracted. To "calibrate" the spectroscope—that is, to determine what wave length of light comes out at what angle—you can measure the angles of some easily generated spectral lines by holding compounds containing various atomic species in the alcohol flame. Their wave lengths have been measured with spectrometers, which use accurately ruled "gratings" instead of prisms. Among the best atomic species for this purpose are:

Species of atom	Wave length in millimeters
Potassium ("alpha" line, red)	0.000768
Lithium (red)	0.000671
Sodium (yellow)	0.000589
Thallium (green)	0.000535
Strontium (blue)	0.000461
Potassium ("gamma" line, violet)	0.000405

Plot the observed angle for these lines against wave length and draw a smooth curve through the points, to get the wave length corresponding to other observed angles.

TABLE OF CRYSTAL CLASSES

The thirty-two crystal classes can be exemplified in the way that Chapter XIII describes, with the cardboard models shown below. Draw the patterns on thin cardboard and mark the "striations" while the pattern is flat. Fold the pattern so that the markings are on the outside and fasten the edges with adhesive tape, or with tabs left on the pattern as in Figure 61.

CUBIC

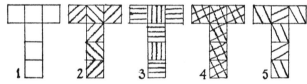

1. Three 4-fold axes through opposite faces
   Four 3-fold axes through opposite corners
   Six 2-fold axes through opposite edges
   Three planes parallel to faces, called the "cubic" planes
   Six planes through opposite edges, called the "dodecahedral" planes
   Center of symmetry

2. Four 3-fold axes, three 2-fold axes through opposite faces, the six dodecahedral planes

3. Four 3-fold axes, three 2-fold axes through opposite faces, the three cubic planes, center of symmetry

4. Three 4-fold axes, four 3-fold axes, six 2-fold axes through opposite edges

5. Four 3-fold axes, three 2-fold axes through opposite faces

TETRAGONAL

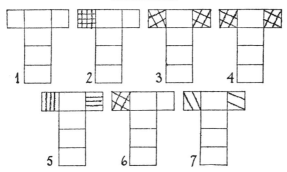

1. One 4-fold axis through the two square faces

   Four 2-fold axes perpendicular to the 4-fold axis, two of them through opposite faces and the other two through opposite edges

   Five planes: (a) one perpendicular to the 4-fold axis, (b) two containing that axis and parallel to opposite faces, (c) two containing that axis and passing through opposite edges

   Center of symmetry

2. One 4-fold axis, four planes containing that axis

3. One 4-fold axis, one plane perpendicular to that axis

4. One 4-fold axis, four 2-fold axes

5. The 4-fold axis becomes a 2-fold axis, of a special type called a "4-fold alternating axis." Turning the crystal through a quarter turn about an axis, and then reflecting it in the plane perpendicular to that axis, is the symmetry operation corresponding to a 4-fold alternating axis (see footnote p. 288).

   Two 2-fold axes, perpendicular to the 4-fold alternating axis, through opposite edges

   Two planes containing the 4-fold alternating axis, parallel to opposite faces

6. One 4-fold axis
7. One 4-fold alternating axis (see footnote p. 288)

ORTHORHOMBIC

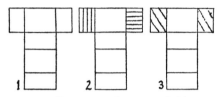

1. Three mutually perpendicular 2-fold axes through opposite faces
   Three planes parallel to opposite faces
   Center of symmetry
2. One 2-fold axis, two planes containing that axis
3. Three 2-fold axes

MONOCLINIC

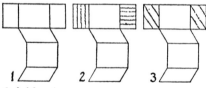

1. One 2-fold axis
   One plane perpendicular to that axis
   Center of symmetry
2. One plane
3. One 2-fold axis

TRICLINIC

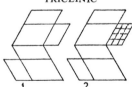

1. Center of symmetry
2. No symmetry

HEXAGONAL

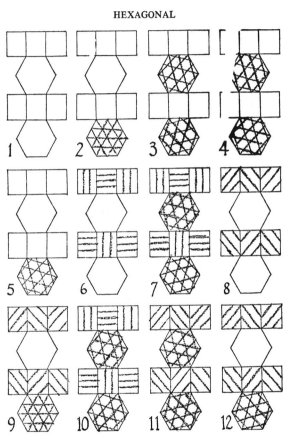

1. One 6-fold axis
   Six 2-fold axes perpendicular to the 6-fold axis,
      three of them through opposite faces and three
      through opposite edges
   Seven planes, (a) one perpendicular to the 6-fold

axis, (b) three containing that axis and passing through opposite faces, (c) three containing that axis and passing through opposite edges
Center of symmetry

2. One 6-fold axis, six planes containing that axis

3. One 6-fold axis, one plane perpendicular to that axis, center of symmetry

4. One 6-fold axis, six 2-fold axes

5. One 6-fold axis

6. One 3-fold axis, three 2-fold axes through opposite faces, three planes through opposite faces, one plane perpendicular to the 3-fold axis

7. One 3-fold axis, one plane perpendicular to that axis

8. One 3-fold axis which is a 6-fold alternating axis.* A rotation of one-sixth turn, followed by reflection through a plane perpendicular to the axis, is the symmetry operation.
Three 2-fold axes passing through opposite faces
Three planes passing through opposite edges
Center of symmetry

9. One 3-fold axis, three planes through opposite edges

10. One 3-fold axis, three 2-fold axes through opposite faces

11. One 6-fold alternating axis,* center of symmetry

12. One 3-fold axis

* For technical reasons, crystallographers have recently agreed to switch from "alternating axes" to "inversion axes"; that is, the operation of rotating the object about an axis and then inverting it through a center on the axis, instead of reflecting it in a plane perpendicular to the axis. You might be interested to check that a 4-fold alternating axis is equivalent to a 4-fold inversion axis, and a 6-fold alternating axis is equivalent to a 3-fold inversion axis.

SUGGESTIONS FOR RESEARCH

The following suggestions for small research projects are genuine: the authors believe that the suggested work has not been done, and that the results of doing it would not be entirely trivial. But the authors have not searched the literature exhaustively for previous work.

1

The crystals of calcite found in nature vary greatly in habit; faces different from the cleavage rhombohedron's often dominate their shapes. On the other hand, sodium nitrate, although it has the same structure, normally shows only cleavage rhombohedron faces when it is grown from a water solution. Some changes have been noticed when the solution contains ammonium nitrate or sodium silicate, but little so far has been accomplished in changing sodium nitrate habit by adding impurities to the growing solution. A distinct habit change would be of interest.

2

Concentrated sulfuric acid produces good etch pits on calcite cleavage faces, and they clearly show the symmetry of the crystal, but this solvent does not work well on sodium nitrate. When you etch calcite with the acid, the product is calcium sulfate, which has a low solubility—the etching proceeds slowly enough to give you time for study of the pits. Further, since calcite is insoluble in water, you can wash the acid off. But you cannot wash sodium nitrate in water, and since the etching product is soluble, the process goes too fast. It would be of interest to find a good sodium nitrate etching procedure with noncorrosive fluids.

3

Many copper compounds crystallize in atomic arrangements in which zinc, nickel, cobalt, iron, manganese or magnesium can replace the copper. In many others the copper apparently cannot be replaced. So far, no crystal similar to calcium copper acetate hexahydrate but containing other elements than copper has been reported. Try to grow such a crystal, experimenting over a wide range of temperature conditions and of calcium concentration in the solution.

4

When a salt hydrate loses water, the loss must of course begin at a surface of the crystal, which turns a powdery white. Dehydration usually begins at a few isolated spots on the surface, and spreads from those spots. It would be of interest to examine the form of this spreading, to see in particular whether the spots take a shape which indicates the symmetry of the crystal, as etch figures do. Strontium formate dihydrate might be a good candidate on which to examine such "dehydration figures."

5

Many tartrates grow quite good crystals. For example, tartaric acid, ammonium tartrate, potassium tartrate hemihydrate, and potassium lithium tartrate monohydrate. Since copper compounds make such large habit changes in Rochelle Salt, it would be of interest to see whether they affect the habit of crystals of other tartrates. If so, the effect can perhaps be ascribed in part to the

presence of the tartrate ion; if not, the effect must be
specific to the atomic arrangement in Rochelle Salt.

6

Since the process of gliding a crystal of sodium nitrate
produces a twin crystal, it may be possible to use a piece
of sodium nitrate which has both glided and unglided
parts as a seed from which to grow a larger twinned
crystal. This would require careful technique; the product—a large clear twinned crystal of sodium nitrate—
would be an interesting exhibition piece.

7

You can guess that the randomness with which right-
handed and left-handed seeds of sodium chlorate form
in a solution might be upset by suitable influences. In
other words, under some influence, right or left seeds
might predominate. For example, if the seeds formed in
some solvent which itself was right- or left-handed (like
citronella), this upset might occur. In fact a slight influence of this kind has been observed in water solutions
of sodium chlorate which also contain dissolved dex-
trose, whose molecules are right-handed. There is another line of attack, which has apparently not been in-
vestigated. If seeds are induced to form on fragments
of some other solid which is right- or left-handed, they
might show a preference in their own handedness. In
such a search, you must be careful that the seeds you
finally examine are not all formed from a single seed
which has been knocked about in the solution, producing more of its own type.

8

When crystals of sodium chlorate grow from a solution
containing borax, their normally cubic habit is changed

to a tetrahedral habit. Probably borate ions are selectively adsorbed on the tetrahedron surfaces, slowing their rate of growth sufficiently to make them appear. On the other hand, sodium bromate, which has the same crystal structure as sodium chlorate, normally grows in tetrahedra without the addition of borax. The corners of the sodium bromate tetrahedra are usually cut off by little triangular faces which belong to another tetrahedron. It is not known whether the major or the minor faces on sodium bromate correspond structurally with the tetrahedron faces borax produces on sodium chlorate. The question might be settled by growing sodium bromate crystals from a solution containing borax, to see whether the minor faces disappear or become larger.

9

Since sodium bromate has the same crystal structure as sodium chlorate, it also rotates the plane of polarization of plane-polarized light. In the case of sodium chlorate, the amount of rotation is known to vary with the wave length of the light, and presumably there is a similar variation in sodium bromate. It would be interesting to measure this variation in sodium bromate at several wave lengths, and plot a curve of rotation against wave length.

The tetrahedra of sodium bromate are not as convenient to work with as the cubes of sodium chlorate; but each major face is usually accompanied by a parallel minor face, providing surfaces between which there is a constant length of path for the light. Sources of light of several different wave lengths are described in the discussion of the spectroscope in the Appendix. It would be wise to practice with sodium chlorate until you can duplicate the measurements previously made by others. You will probably find that you must use a spectroscope to isolate the color of light with which you wish to work.

10

The recipe for growing calcium copper acetate points out that the crystal forms from solution only when the proportion of calcium to copper in the solution is greater than the proportion in the crystal. It would be of interest to find the minimum relative calcium concentration at which calcium copper acetate is deposited instead of copper acetate monohydrate. In general you can expect that the critical relative concentration will vary with temperature.

BOOKS AND ARTICLES

There are not many books or articles on crystals easily found in many libraries or at a level similar to that of this book, but those that do exist are of remarkably high quality. The most extensive choices are in the area where crystals, geometry, symmetry, and art touch upon each other. For instance, much of the work of the Dutch artist Maurits Escher has touched this theme. *The Magic Mirror of M. C. Escher* by his friend the physicist Bruno Ernst (Random House, New York, 1976) explores the way in which Escher developed his space-filling pictures. The crystallographer Caroline MacGillavray treats Escher's work as precisely crystallographic in *Fantasy & Symmetry: the Periodic Drawings of M. C. Escher* (Harry N. Abrams, New York, 1976). A similar game is played out by Peter Stevens in his *Handbook of Regular Patterns* (MIT Press, Cambridge, MA, 1980), a field guide to all crystallographic groups possible for 2-dimensional "crystals," illustrated by worldwide examples drawn from the arts of many peoples. A look at symmetry geometry not limited by the crystallographic need for arrayed repetition can be found in *Shapes, Space, and Symmetry* by Alan Holden (Columbia University Press, New York, 1971), while some of the crystallographic properties to be noticed in woven cloth are explored in *Spiders' Games, a Book for Beginning Weavers* by Phylis Morrison (University of Washington Press, Seattle, WA, 1979). (*Crystal*'s authors, too, could not lay the subject aside.) All of these broad views of crystals are returned with vast richness to the realm of science in Hermann Weyl's exploration of the profound ideas of orderliness in nature and human art, *Symmetry* (Princeton University Press, Princeton, NJ, 1952).

The *Scientific American* magazine is a good place to search for up-to-date articles connecting crystallography with other areas of science. For instance, the issue of September 1977 is devoted to microelectronics, and has many references to crystals, especially the article by James L. Meindl, "Microelectronic Circuit Elements." Elizabeth Wood's *Crystals and Light* (Dover Publications, New York, 1977) touches on some crystal physics, while the chemists Bernal, Hamilton, and Ricci's *Symmetry, a Stereoscopic Guide for Chemists* (W. H. Freeman, San Francisco, 1972) is a guide to the shapes of molecules in three dimensions, complete with a stereo viewer and stereo-pair drawings of the crystallographic point groups. Any good field guide to rocks and minerals will deal with geology and crystals in nature. The inexpensive Golden Nature Guide *Rocks and Minerals* by H. S. Zim and P. R. Shaffer (Simon & Schuster, New York, 1957) has exceptionally well made color paintings of crystals and minerals: they are quite good enough to help the amateur learn to identify new minerals, either in the field, or in museums. In *Gems Made by Man* (Chilton Book Company, Radnor, PA, 1980), Kurt Nassau studies the central subject of this book, showing how diamonds, rubies, sapphires, emeralds, and more were first grown in the laboratory and how they are grown today. There is a particularly interesting chapter on understanding the color of gem crystals.

November 1981

ANSWERS TO PROBLEMS

PROBLEM 1

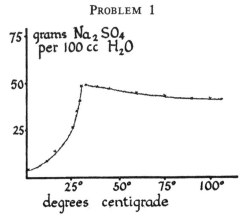

The solubility of sodium sulfate decahydrate increases with temperature, while the solubility of anhydrous sodium sulfate decreases with temperature. The temperature at which the solution is saturated with both the hydrated and the anhydrous salt is 32.5°C. Above 32.5° the solubility refers to a solution saturated with the anhydrous salt. Below 32.5° the solubility refers to a solution saturated with the decahydrated salt.

PROBLEM 2

A. The Sealed Jar Method

a. Saturating the solution at the growing temperature.
b. Heating the solution above the growing temperature preparatory to dissolving more salt in it.
c. Dissolving additional salt in the solution at the higher temperature.

d. Cooling the solution to the growing temperature after it has been seeded.

e. Allowing the crystal to withdraw material from the solution at the growing temperature.

B. The Evaporation Method

a. Saturating the solution at the growing temperature.

b. Heating the solution so that it can be seeded while it is unsaturated.

c. Cooling the solution to the growing temperature after it has been seeded.

d. Allowing the solution to evaporate at constant temperature, becoming supersaturated while the crystal grows.

PROBLEM 3

There is danger of depositing the decahydrate at temperatures below 32.5°C., and a growing temperature above that should be picked. Because the solubility of anhydrous sodium sulfate decreases with increasing temperature, the supersaturated solution must be made by first saturating a solution *below* the growing temperature, and then heating it to the growing temperature to supersaturate it after seeding. But since the solubility of the decahydrate behaves normally, the initial saturated solution must be prepared at a high enough temperature for it to yield a supersaturated solution when it is heated.

PROBLEM 4

If you have grown crystals of both the heptahydrate and the hexahydrate of nickel sulfate at room temperature, you could mix the two solutions from which you grew the different crystals and hang a seed of each of the hydrates in the mixed solution. This solution would be represented by a point lying between the two solubility curves. The hexahydrate seed should dissolve at the same time that the heptahydrate seed grows, so long as the temperature stays fairly constant at your original growing temperature.

PROBLEM 5

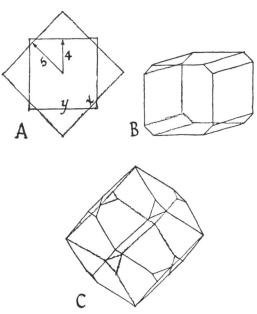

A. The crystal will look like an octagon, whose y-faces are larger than its x-faces.
B. The crystal will be a cylinder, whose cross-section has the shape of the octagon of crystal A, terminating in faces perpendicular to the axis of the cylinder. The length of the cylinder will be twice its width.
C. The crystal will look like a rhombic dodecahedron, eight of whose corners are cut off by little triangular faces.

PROBLEM 6

1. F, G, J, L, P, Q, R
2. N, S, Z
3. A, M, T, U, V, W, Y
4. B, C, D, E, K
5. H, I, O, X

PROBLEM 7

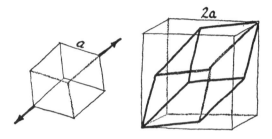

1. The face-centered cubic structure can be built of rhombohedral unit cells with atoms only at the corners of each rhombohedron (Figure 92). Since the rhombohedron can be regarded as a cube that has been stretched along a diagonal from corner to opposite corner, you can consider the face-centered cubic structure as obtainable by stretching the simple cubic structure to the right extent. Notice that the corners of the rhombohedron in the diagram will fall in the right places if the cube is stretched so that the diagonal from corner to opposite corner is doubled in length. In other words, a simple cubic arrangement is converted to a face-centered cubic arrangement by stretching it along an axis of three-fold symmetry until every spacing along that axis is doubled, and spacings perpendicular to that axis remain unchanged.
2. Reasoning in the same way with the rhombohedral

unit cell for the body-centered cubic structure (Figure 95), you find that it can be converted to the simple cubic structure by stretching it along an axis of three-fold symmetry until every spacing along that axis is doubled, and spacings perpendicular to the axis remain unchanged.

3. Since you have just shown that the face-centered cubic structure can be converted to the simple cubic structure by squashing it along an axis of three-fold symmetry, and that the simple cubic structure can be converted to the body-centered cubic structure by squashing it along the same axis still further, you can convert the first structure to the last by squashing it along an axis of three-fold symmetry until it is one quarter of its original length.

In Figure 91 you see that the body-centered tetragonal building block for the face-centered cubic structure can be converted to a body-centered cubic building block (Figure 94) by squashing it along its axis of four-fold symmetry until it is $\dfrac{1}{\sqrt{2}}$ times its original length.

PROBLEM 8

1. Both of the structures have two planes of symmetry. In structure B, the electrically neutral planes, along which the crystal may cleave easily, are parallel to the planes of symmetry; in structure A they are not. Hence, in structure A the two families of cleavage planes are equivalent, because the planes of symmetry reflect each family into the other. In structure B the two families are different, as in calcium copper acetate hexahydrate (see Plate 36); and you would expect cleavage to be better in direction #1 than in direction #2 because the planes parallel to direction #1 are farther apart. Outlining a part of the

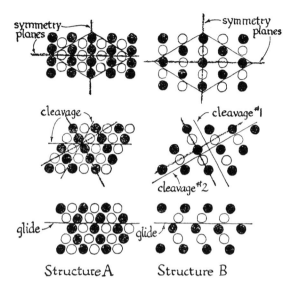

Structure A   Structure B

structure with a rhombus makes the directions of possible glide especially clear, and shows that the glide planes and the cleavage planes coincide in structure A and do not in structure B.

2. The cubic unit cell for diamond (A) shows that four bonds must be broken within each cell in order to cleave diamond parallel to the cube faces. Calling the length of the side of the cell a, you can write the number of bonds broken per unit area as $4/a^2$.

Another way of looking at this counting problem, more closely comparable with the way you must use in looking at cleavage along the octahedron faces, is as follows. Each atom in any cube plane has two bonds to atoms in the adjacent plane. Thus, the number of bonds to be broken is twice the number of atoms in a plane. If you regard the face of the unit cell as a two-dimensional building block, you see that the block contains two atoms and has the area a^2. Hence, the area per atom is

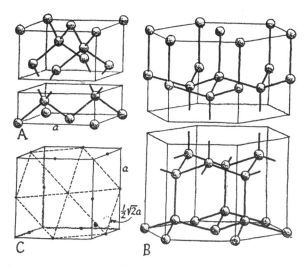

$\dfrac{a^2}{2}$, the area per bond is $\dfrac{a^2}{4}$, and the number of bonds per unit area is the reciprocal of that last figure, or $4/a^2$.

In order to find the number of bonds broken per unit area in cleaving along octahedron faces, notice that the diamond structure can be regarded as consisting of two interpenetrating face-centered cubic structures, displaced with respect to each other in the direction perpendicular to the octahedron faces. The amount of that displacement is one quarter of the length of the body-diagonal of the cubic block. Drawing the structure from the point of view taken in stacking marbles in the face-centered cubic structure (B), you will see that there are octahedral planes crossed by one bond per atom.

The problem of the number of bonds to be broken in cleavage then becomes the problem of determining the area of a two-dimensional hexagonal building block and the number of atoms it contains. The number of atoms has been shown to be three. Reference to Figure 90C shows you that the side of the hexagon (shown again at C) has the length $\frac{1}{2}\sqrt{2}a$. Then the area of the hexagon is $\frac{3}{4}\sqrt{3}a^2$, the area per bond is $\frac{1}{4}\sqrt{3}a^2$, and the number of bonds per unit area is $\frac{4}{\sqrt{3}a^2}$. Hence, the number of bonds which must be broken per unit area is less in octahedral cleavage than in cubic cleavage, by the factor $\frac{1}{\sqrt{3}}$, which is a little more than one half. In fact diamond shows very good octahedral cleavage, and the cleavage is often a help to the gem cutter.

PROBLEM 9

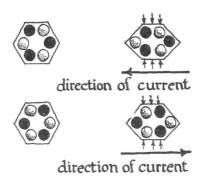

direction of current

direction of current

1. A squeeze perpendicular to the "a" faces of the two-dimensional crystal would produce a current *perpendicular* to the direction of the squeeze, in contrast to the current *along* the direction of the squeeze which is produced when the squeeze is perpendicular to the "b" and "c" faces. Thus this electrical effect is quite sensitive to direction.

2. The diagram of current directions for different squeeze directions has the symmetry of the structure, and no higher symmetry. Thus, measurements of the piezoelectric effect could be used to reveal the true symmetry, but directional piezoelectric experiments

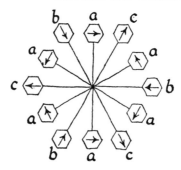

are quite difficult. A test for the *existence* of a piezo-electric effect in a crystal, however, is often used as a test for the *absence of a center of symmetry*.

PROBLEM 10

The models have the symmetries described in the table of crystal classes as:

A. Hexagonal 4
B. Hexagonal 10
C. Hexagonal 9
D. Hexagonal 12
E. Tetragonal 4
F. Tetragonal 5

INDEX